EARTH'S
CHANGING CLIMATE

By the same author

NOISE POLLUTION
LONDON'S DROWNING
FLOODSHOCK
OUR DROWNING WORLD

EARTH'S
CHANGING CLIMATE

The Cosmic Connection

Antony Milne

PRISM
PRESS

Published in Great Britain in 1989 by
PRISM PRESS
2 South Street
Bridport
Dorset DT6 3NQ

and distributed in the USA by
AVERY PUBLISHING GROUP INC.
350 Thorens ASvenue
Garden City Park
New York 11040

ISBN 1 85327 040 7

Typeset by Maggie Spooner Typesetting, London
Printed and bound in Great Britain by Biddles Ltd,
Guildford and King's Lynn.

CONTENTS

For all the mysteries our solar system still contains, these can be nothing compared with those of the universe as a whole. Our little corner of space is not a closed system; there is no fence around it to keep trespassers out, and much of the energy that sweeps around our planet originates from far beyond its outer limits.

The Cycles of Heaven,
Guy Lyon Playfair and Scott Hill

INTRODUCTION

Goesta Wollin, of the Lamont-Doherty Geological Observatory of Columbia University, New York State, has in recent years made a name for himself by predicting severe inclement weather. For some while he has been convinced that firm predictors of violent storms and blizzards in America are sudden changes in the strength of the solar magnetic field, in which Earth's field is embedded.

Wollin, increasingly convinced of the magnetic link, began, on 22 January 1986, to phone tv stations in northeastern parts of the US. He was mindful of the publicised prediction he made about two months earlier of a severe storm in the area, which took the lives of thirty people and which the forecasters had failed to anticipate. This time he pointed out that a colleague at the Fredricksburg Magnetic Observatory in Virginia had reported a marked change in the magnetic flux of about 45 gamma. This, said Wollin decisively, meant there would be a major snow-storm or flood within the next six days in the region. He pleaded that his predictions be mentioned on the air alongside those of the weather forecasters. They showed no interest.

However, three days later Wollin's premonitions were justified, with a metre of snow falling in New England to the north of Boston, and several inches of rainfall over the coastal region from Boston to Washington, with the loss of several lives and much property.

It was not just Wollin who was vindicated, but the painstaking work of seventeen US universities, together with seven federal agencies taking part in a $10 million study known as GALE (Genesis of Atlantic Lows Experiment), using data from ships, ground stations and research satellites. The primary goal of GALE is to test and improve computer-based weather forecasting of storms in the eastern seaboard area, a particularly densely populated region.

1

The new science of 'magnetometeorology' has, in fact, a long pedigree and is supported by a wealth of data from distinguished centres like the Aspen Institute and the National Centre for Atmospheric Research, both in Colorado. Even so, the last decade has been a fast-moving one for American climatologists like Wollin. New knowledge about how sunspots *can* affect the weather, about the way the Sun and natural convective processes can actually deplete the ozone layer much more effectively than can mankind and about the way ancient climates may be linked to lunar and planetary cycles, is constantly emerging.

In addition this book will be looking at the stunning new theories that have come out of the University of Berkeley, California, which endeavour to prove that a giant asteroid struck the Earth 65 million years ago and possibly brought about the death of the dinosaurs. These theories, first advanced in 1980, are withstanding scientific criticism well and could, within a few years, completely revise our theories of evolution and extinction. They have stimulated new discussions about how the climate and biosphere could be affected by such a massive impact.

Could such a missile land in the near future to play havoc with the world's weather? Scientists are still trying to work out the answers. But what is clear is that, with the help of computer modelling, a new and unified extraterrestrial dimension to the study of climatic change is at last emerging.

PART I:
THE COSMIC CONNECTION

Chapter 1:
THE COSMIC CAROUSEL

Since 1945 Man's awareness of his place in the cosmos has gathered revolutionary momentum. It now goes beyond the dimly perceived understanding that planet Earth is merely a tiny cog in a massive cosmic carousel. It is the notion, widely held, that real, tangible phenomena in outer space have actually affected Earth in direct and sometimes catastrophic ways.

For the truth is, the universe is dynamic: everything in it from particles to nebulae is in a constant state of flux and movement. Since you started reading this sentence the Earth has travelled 100 miles around the Sun, the Sun has moved 1,000 miles in its orbit in the galaxy, and the orion nebula has moved 100,000 miles relative to us. Stars are born or explode violently in death throes, meteorites collide with moons and magnetic fields affect the behaviour of living organisms. Energy can transform itself into heat and can — even across unfathomable distances — determine varitions in heat and cold, both on Earth and elsewhere, through gravitational, electromagnetic, or kinetic energy forces. Indeed, universal forces succeed in regulating literally trillions of energy cycles in all organic and inorganic matter, throughout space. They are all miraculously harmonized at the stellar, planetary, climatic and even biological level into one everlasting and deep 'song' of life, a concept first envisaged by Copernicus and Kepler.

The word 'miraculous' is used advisedly, since there is now much talk about the universe being intelligent, showing a construction virtually indistinguishable from that of a giant computer. The idea is rooted in the fact that most of the universe consists of pure electromagnetic energy. In radio waves we use this energy as a form of data transmission. Cosmic rays, of which we will have more to say later, may themselves be described as a form of data communication. This is where the computer analogy comes in. A computer arranges an electronic pattern (the program), which interacts with other electronic patterns (the data memories) to produce new patterns (data processing).

The analogy can be pushed further. Data processing is how life is organized at the molecular level. Take a plant or animal. Molecular biology grew out of earlier discoveries which proved it was the genes that determined the structure of living creatures. Later chemical analysis revealed that the DNA molecules, which had a pattern of atoms that make virtually a 'blueprint for life', contain all the data to determine growth, health and species differentiation. In other words, the cosmos is continually disseminating information codes across a variety of spectra, codes ultimately responsible for the make-up of all matter and biological life.

Let us look for a moment at the way these information codes in the form of harmonized universal forces affect the temperatures and chemical make-up of the solar planets. Why, we might ask, are the four rocky planets those that are closest to the Sun? Common sense tells us that the planets nearest the Sun should be the warmest, and those furthest away the coldest. But this does not itself explain why the four planets furthest away from the Sun are giant balls of frozen gases and ice. A more scientific answer would explain that some mineral grains, such as the metals and silicates, could condense out at quite high temperatures, as would apply to those parts of the dust cloud nearest the embryonic Sun. These gritty grains eventually solidified to form Mercury, Venus, Earth and Mars. Hence the remaining condensing gaseous substances, now much further away from the Sun, froze into incredibly massive balls of ice molecules. They became Jupiter, Saturn, Uranus and Neptune (we will ignore Pluto, which has an odd mode of rotation, is very small, and is probably an escaped moon). In this way the actual characteristics of the solar planets were determined at birth by the nature of the mineral elements in the primordial warm dust cloud from which they first came. Many of these elements were in fact second-hand (i.e. derived from the spent plasma of dying, exploding stars). The borrowed atoms in the centre of the gas cloud were squeezed into the centre by gravity to form a dense, scorching mass. As temperatures rocketed, nuclear reactions erupted violently, and the Sun was born in a blaze of light. From the cooler outer reaches of this cosmic debris the planets were created.

As the primeval cloud cooled further, each crystalized molecule

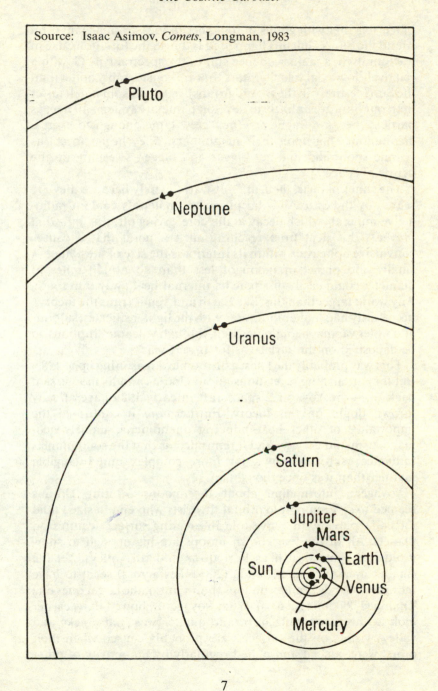

Source: Isaac Asimov, *Comets*, Longman, 1983

Pluto

Neptune

Uranus

Saturn

Jupiter
Mars

Sun

Earth

Venus

Mercury

of ice, together with grains of rock, began to swirl in a giant orbit round the Sun. (That this happened is due to the fortuitous size of the Sun itself, a main sequence star of the spectral type G, with a relatively low axial velocity. Stars that are considerably hotter than the Sun (i.e. those in the B or A groups), or which spin much faster than our Sun, are unlikely to develop planetary systems.) Then, as particles became fragments, they grew large enough to exert a gravitational tug upon their neighbours. They began to act as cosmic snowballs, and got bigger and bigger, sweeping up the debris of space.

The inner planets, then, are not warm merely because they are seared by the Sun. All solid planetary materials contain radio-active minerals which decay at the core, giving off heat, but what prevents this heat from reaching such a point that it causes convulsive upheavals within its interior is the size of the planet. A small solid planet or moon of less than about 150 miles in diameter could easily dissipate its internal heat away into space. Any world larger than this, like Earth and Venus (runs the theory), necessarily had molten-hot cores with the lighter silicates, sulphur and water vapour, being 'outgassed' through volcanic eruptions, to be deposited on the surface as lavas and seas.

This was probably the latest astronomical reasoning up to 1985. But the astonishing revelations about Uranus and its moons sent back from the Voyager 2 spacecraft in early 1986 have already thrown doubt on this theory. Furthermore it confirmed the importance of other heat-inducing phenomena; namely geo-magnetism and gravity. And it reminded us that the solar planets influence each other in a far more complex and intangible manner than was once thought.

Voyager's information about the moons orbiting Uranus seemed to give support to orbital theorists who emphasized tidal forces as playing an active role in creating surface features on planets. All of the five known moons are big enough to show geological activity such as lava flows and rift valleys. Yet the smaller ones such as Miranda (300 miles across) seem to have more cracks, ridges and striations than some of the larger ones like Titania (1,000 miles across). What Voyager 2 showed in its earlier look at the moons of Saturn and Jupiter was that cracks and valleys visible on the surface were probably caused when their orbits were pushed and pulled regularly by the gravity of other satellites. This, according to some astronomers, was what brought about the 'stretch marks' on Jupiter's moon Ganymede, and on Saturn's moon Enceladus. In addition, Stan Dermott, of Cornell University, has calculated that there could have been episodes when internal heat was the product of mutual gravitational attraction between Uranus's moons.

Richard J. Greenberg and Robert L. Marcialis of the University of Arizona's Lunar and Planetary Laboratory also suggested that tidal heating may have warmed Miranda, as well as Jupiter's moon Io. In fact the Voyager spacecraft discovered active volcanoes on Io, with measurements of infrared radiation suggesting that the moon emanates 100,000 gigawatts of power — and one gigawatt of power is the normal output of one nuclear power station. Now Greenberg and colleagues believe they will be able to check back into Io's historical orbit to see if strong tidal forces have altered its orbits since Gallileo first spotted its existence in 1610. It is already known that Io orbits Jupiter at twice the rate of Europa, which in turn orbits the planet at twice the speed of Ganymede. This harmonious relationship means that every now and then they return to the same orbital positions. And almost certainly Io is absorbing energy within a sophisticated resonating relationship from Europa and Ganymede. It is this tidal force 'resonance' that is said to heat Io sufficiently to cause volcanoes to erupt, through complicated geochemical processes involving the smelting of sulphur deep inside the moon.

However, not all scientists agree with the tidal heating theory. We have seen that the outer solar planets are extremely cold and gaseous, and it would not be surprising to find their moons also to be icy. Strangely, Steven K. Croft of Arizona University says that Miranda could have warmed itself if the ice at its core was composed partly of frozen methane as well as water-ice. This kind of ice is known as clathrates. Laboratory models show that water molecules in the clathrates can trap methane, carbon monoxide and other particles. This makes them excellent insulators, capable of capturing heat generated by decaying internal radioactive isotopes. So after about a million years, says Croft, the clathrate ice could remain solid, while melting other ice on the moon consisting of ammonia and water-ice compounds. This melting process would be enough to expand the core by one per cent, enough to crack the face of the moon.

9

Relaunching the Jupiter Effect

However, the tidal theory seemed to be the most likely, and Voyager appears to be giving new credence to a recent theory frequently dismissed by scientists, known as the 'Jupiter Effect'. This was a widely publicised gravity-based doomsday theory circulating in the early 1970s and derived from Chinese astronomical research. The theory dictated that the combined tidal forces of the solar planets have an electromagnetic effect on the Sun, and can cause violent sunspots and flares that in turn can precipitate climatic change here on Earth. The authors of a seminal book, *The Jupiter Effect*, were John Gribbin and Stephen Plagemann, both Cambridge astrophysicists involved in some of the early 1970s research in cosmology.

In order to better understand how the Jupiter Effect works, we need first to have a more detailed understanding of gravity. In an important sense the component parts of the universe are interrelated through four universal forces: electromagnetism, the weak and the strong nuclear forces that apply to particles and their constituents, and gravity. So far the power of gravity seems the odd man out, since it is far, far weaker than the other three forces. The 6,000 million million million tons of Earth can still produce only a modest gravity keeping us bound to its surface. In fact we can generate magnetic or electric forces hundreds of times more powerful with just a few pounds of magnetised iron. It has been calculated that the gravity waves emanating from the entire planet have an energy of about a millionth of a horsepower, and the total emission from the solar planets and the sun together is only half a horsepower. Yet, being cumulative in power, it dominates on the grand, cosmic scale.

According to physicist Paul Davies of Newcastle University, the story of the universe is really one of 'the struggle against gravity'. Not only do bodies pull on each other, they tug inwards continually to compress themselves into smaller objects. Similarly, the squeezing gravitational force within the tiny frame of a human being is overwhelmed by the powerful downward attraction of Earth's gravity. Out in the cosmos, the gravity emanating from the core of a star prevents its gases from forever trying to revert to the simpler, formless state existing in the primordial universe.

Gravity, of course, is vitally important for us here on Earth. It prevents our atmosphere from escaping back into the space from which its atoms originated, it affects the tides of the oceans, and literally holds all solid matter together. Equally importantly it generates life-sustaining energy through the nuclear fusion process taking place in the Sun. All this is not just theory. We know that gravity exists because of the way objects on Earth and in space have been observed by astronomers to affect each other at a distance. Other experiments prove that every mass-producing object in the universe exerts a gravitational pull on every other object within its field, and vice versa. Thus, depending upon mass and distance, they will accelerate towards each other — the larger and nearer one pulling the other closer to it. So the Sun pulls on Mercury (the planet closest to the Sun) more strongly than it pulls on Venus or Earth.

Scientists have still not managed to utilize gravity or invent anti-gravity devices of science-fiction fame, but some theoretical physicists believe they now have a much better insight into the nature of the force. They speculate that there may be a fifth and even a sixth universal force, the latter strengthening gravity and the former weakening it. Experiments performed on top of a 600-metre tv tower in North Carolina seem to prove that the so-called sixth force added five parts in 100 million to gravity's attraction. Frank Stacey and his colleagues at the University of Queensland, writing in December 1987, suggest that the fifth force may in fact be the difference between two much larger forces of repulsion and attraction. Other scientists say that a short-range repulsion could be a component of gravity itself. It is fair to point out however, that recent studies of planetary orbits, and other more mundane experiments on Earth with bits of copper and tungsten wire, cast doubt on the fifth force theory. Nevertheless the original proponents still argue that it does exist, but at a limited, terrestial range.

The discovery of the Black Hole in the early 1970s was to confirm the devastating force of gravity; it also illustrated how the powers of nature were akin to magic, by the way matter with immense mass could be squeezed down to 'nothingness'. Yet any void resulting from this crushing force could still hide incredible and fearsome phenomena that destroys in a split second any matter (even a planet) that approaches it. One, perhaps several, of these Black Holes may exist in our own galaxy. Paul Ho and his colleagues from the Harvard–Smithsonian Centre for Astrophysics have spotted a streamer of cool, barely visible gas in the process of falling into the centre of the galaxy, very likely, he suggests, because a Black Hole lurks there.

A Black Hole might miss the Sun in its travels, but will be greatly affected by the Sun's gravitational field, or even be captured by the Sun. Gravitation pulls and stretches, and a close encounter with such an awesome intruder would at the least hump up Earth's oceans to cause devastating floods around the world. It could even disturb the Earth's orbit — enough to dramatically alter the climate. Even a mini Black Hole, likely to be of asteroidal proportions, with a mass one-millionth of the Earth, can be a highly dangerous phenomenon. In the event of a direct hit a mini Black Hole would tunnel its way right through the mantle of the Earth, consuming and vapourizing matter on its journey to emerge, enlarged, from the other side. As the Black Hole reached the crust, however, an enormous explosion would occur, devastating a considerable area of the biomass and dangerously atomizing the atmosphere. It was suggested that this caused the destruction of Tunguska, Siberia, although as no exit point (with similar explosive repercussions) occurred on the other side of the globe, this theory has now been discounted.

However, a smaller Black Hole, the size of a pinhead, might be sufficiently slowed by gravity on its journey through Earth and remain lodged in the centre. It could then act as a macabre planetary cancer, slowly eating out the core of the Earth until the planet collapses inward. Finally it would leave only its ghostly gravitational pull as evidence of its previous existence. Unlike the Tunguska Black Hole, this theory has not yet been discredited and could yet explain the increasing number of geophysical disturbances Earth has suffered from in recent years.

Black Holes are not to be found everywhere in space, as they only await the fate of stars several times bigger than the Sun. Although gravity holds stars together, it also determines such things as the pressure inside a star, which prevents it from collapsing in on itself. Energy, on the other hand, is pushed out of the star by electromagnetic radiation. Fortunately billions of lesser, more solid bodies exist in the universe, so gravitational influences will be cancelled out, complemented, reinforced, deflected, and so on. Several asteroids, for example, may be attracted into the orbit of a moon, until disturbed by the stronger pull of a larger object, which causes the velocity of the asteroid to be slowed and ultimately reversed.

Collisions can and do occur in space between hurtling objects sucked into ever-faster orbits. But what prevents self-destructing crashes between orbiting bodies is the orbital motion itself. A centrifugal force is automatically set up which counteracts the inward pull of the larger body, but this 'escaping' force is a form of acceleration, and hence itself a kind of gravity. The stronger the pull, the faster the rate of orbit, and hence the greater the countervailing 'gravity'. For this reason the inner planets rotate considerably faster than the outer ones. The weaker the pull, the more the captured body is able to distance itself and orbit at a more leisurely pace.

The solar planets, of course, exert gravitational pulls on each other. These influences usually cancel each other out, but occasionally the planets are aligned in a distinct synod — not necessarily in a straight line (a rare occurrence) but in a rough

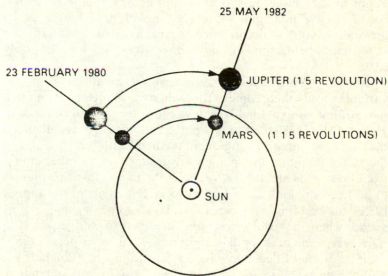

While Mars makes about one and one-fifth orbits, Jupiter makes one-fifth of an orbit. On average, they move into conjunction once every 2¼ years.

Source: *Judgement of Jupiter*, Richard A. Tilms, Book Club Assocs, 1980

semi-circle — and their individual pulling power is greatly enhanced. They all orbit at different speeds: every 2.24 Earth-years Mars and Jupiter catch up with each other, as Mars would have made one and one-fifth of an orbit around the Sun, while Jupiter would have made just one-fifth of an orbit (Jupiter is so far out it takes twelve Earth-years to orbit the Sun); Mercury, moving fastest of all, rotates around the Sun in 88 Earth-days, and would be in conjunction with the other planets every three or four months, each time exerting an additional gravitational tug; the asteroid belt takes about three Earth-years to rotate at its inner edge, but six at its outer edge.

The greater the mass, the greater the pulling power. For instance, the tidal effect of the Earth on the Sun is only 10,000th of that of both the Sun and the Moon on the Earth. Similarly Jupiter, with its radius eleven times that of Earth and with over 300 times its mass, is so much further away from the Sun it exerts a pull that is only 2.3 times greater than Earth. And yet it has a much greater influence over the asteroids, and it can still vie with the Sun in pulling power. Venus has a similar mass to Earth, but being closer to the Sun it has a relative tidal pull some 1.9 times that of Earth; i.e. its gravitational pull is almost twice as great as Earth's. Mars, on the other hand, being smaller and further away, has a pull of only 0.03.

Inevitably in any planetary system with its members orbiting at variable speeds, there comes a time when the planets converge in a more or less straight line. True, it would take tens of thousands of years for all nine planets to be a perfect alignment (it is calculated that the most important alignment is when Earth is on one side of the Sun and all the other planets are grouped on the other), but it would take only about 1,000 years for the more massified planets to be lined up, and, according to some adherents of the Jupiter Effect, the best time for *most* of the planets to be in some kind of synod, when they will crowd together in twos and threes in a narrow segment of sky, will be right now — i.e. in the period 1980–93. If the Jupiter Effect is to have any meaning it must include the pull of Saturn and so, in effect, enable the two to work in tandem. Voyager 2 was helped on its journey from Earth by the aligned pulls of both Saturn and Jupiter. Without the extra gravitational tug the journey would have taken some thirty years; as it was, it took just eight-and-a-half.

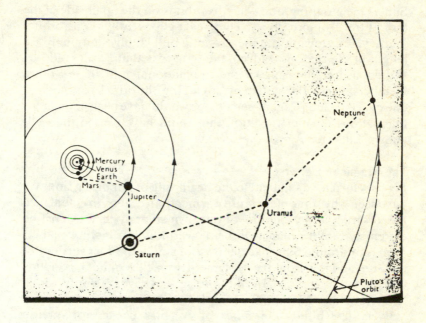

The Jupiter Effect.
This shows the relative positions of the solar planets for the year 1982, when it was predicted that severe seismic activity would take place on Earth.

Source: Patrick Moore, *Countdown*, Michael Joseph, 1983

The essence of the Jupiter Effect, then, is that planets can affect the Sun through their gravitational pulls. Originally the Effect was considered to be so powerful that when the synod was at its peak in 1982, earthquakes could be triggered off around the San Andreas fault. This presentation of Chinese calculations as a new doomsday theory fascinated the world's media, although some critics suggested that the Jupiter Effect was little more than 'fringe' science. This accusation was clearly unfair, since there is plenty of scientific evidence to support such 'forces acting at a distance' theories. Indeed, a dismissal of the Jupiter Effect is a rejection of the very idea of gravity and everything that has been discussed so far. For we know that when Earth becomes involved in a planetary grouping it undergoes an angular deceleration force. It travels more quickly through the half of its orbit on the same side of the Sun as the grouping, and when its orbit is on the other side of the Sun, it travels more slowly. Then, as it glides away on its orbital journey, the growing and then waning pull changes minutely the length of the seasons. Indeed, the danger is that Earth could be shifted nearer the Sun by the gravitational pull of the other members of the solar system by 'stretching' the orbit by as much as one per cent. Such a small percentage would have been enough for fluctuations in solar radiation that could have caused the well-known Little Ice Age of the seventeenth century.

The Magnetic Universe
Let us return for a moment to the beginning of our discussion. We have seen why some planets are warmer than others. In actual fact, based on an understanding of our own solar system, it must be extremely rare for planets to be as warm as Earth since planets form only around Suns like our own, or smaller, colder versions of the Sun, and it is from the cooler, outer reaches of hot gas clouds that planets are formed. In addition, most of space is empty, with a temperature of 2.7 degrees absolute (–270C).

Every few billions of miles or so, tiny pockets of intense superheat can be found, either in the form of stars or as plasma. Plasma is the stuff of which stars are made. It consists of tens of billions of hot atoms, where the gases and dust of space have had their internal particles broken down into positive or negative electrical charges. What causes this to happen is fundamental to the very existence of all life on Earth. If you heat a solid like ice it becomes a liquid, and then a gas. If gas is heated even more it becomes ionized, or plasma, and most of the universe is in this state. We know that all matter, even the thin gases of space, contains electrically-neutral atoms. What binds atoms into molecules is the atom's ability to swap or share its negatively charged electrons. But atoms differ in the number of electrically-charged particles they possess, depending on the type of matter they constitute — gaseous, glass-like, metallic, etc. They are recognised in terms of the electromagnetic interraction taking place between the various subatomic particles. A plasma containing differently charged particles can still be electrically neutral, but can act as a giant conductor of electricity. In a sense

16

any body of ionised gas — with its electrons knocked off — can create an interstellar electric current. The plasma moves violently, and magnetic fields whip through and around the embryonic star from pole to pole. Then, when the star is formed, a net electric effect arises if one or other positive or negative charge predominates — an electric field is created.

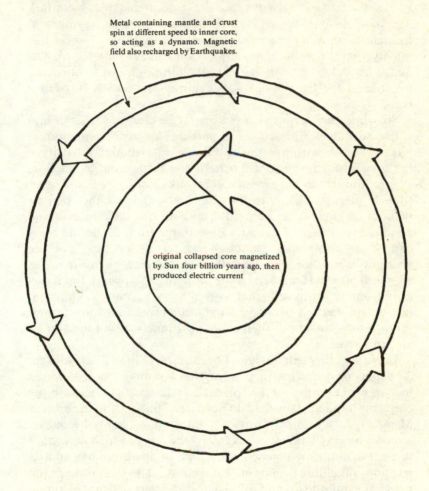

Metal containing mantle and crust spin at different speed to inner core, so acting as a dynamo. Magnetic field also recharged by Earthquakes.

original collapsed core magnetized by Sun four billion years ago, then produced electric current

How Earth's magnetic field is created.

Adapted from: *The Earth*, Granada Publishing, Steve Parker, 1985

We can see, hence, where electromagneticism (EM) gets its name: it is a combination of the forces of electricity and magnetism, in a sense electricity in motion. In 1820 the Danish physicist H.C. Orsted discovered that electricity and magnetism are inextricably linked, and thirteen years later Michael Faraday discovered that a moving magnet can make a current flow through a wire. 'Forces acting at a distance' was interpreted by James Clark Maxwell as a power extending well into space, having real influences on other matter. But EM is a long-range force, like gravity, that acts between material bodies such as matter, planets and stars. Indeed we only see stars at all through EM frequencies, because of the light, radio waves, ultraviolet and X-Rays they emit.

So, while gravity tries to knock aggregate clumps of matter into spherical shapes, EM acts on all particles larger than an atom. It has an overall bearing on the heat flows and radiation intensity of the solar wind, the corpuscular radiation thrown out by the Sun. Planets thought to have molten cores (like Earth), or liquid cores (like Jupiter), are likely to have magnetic fields. Other planets without EM fields may be solid throughout, or lack the essential conducting element, iron. Any EM charge from a celestial body can be detectable upon planetary motions — a compass needle changes its position by about 100th of a degree when the moon rises, but often it is masked by the buffeting solar wind. But it was only from 1954 onwards that astronomers had any notion that planets emitted an EM charge. First Jupiter was discovered to emit radio noises, and from then on other planets were found to be electromagnetic.

There are other telltale signs. For example, Ellis Miner, Voyager 2's deputy project scientist, believes that Uranus's moon Miranda has an effect on the mother planet's magnetic field because the electrons trapped in the field do not extend out beyond the orbit of Miranda. As a result Uranus's satellites are probably heated up not only by gravitational forces, but by the way in which they move through Uranus's magnetic field, which in any event has an axis rotation coinciding with its orbital revolution, so that its magnetic field has a quite distinctive influence. It is just as well that Miranda and Uranus have an established planet/moon relationship, and are not giant careening planetisimals. US physicist Ralph E. Juergens maintains that when two positively charged missiles rendezvous at a point of orbital contact, unleased electrical fields would clash. Almost instantly forces 'immeasurably greater than gravity would be brought to bear on the charged bodies.'

All of these phenomena have, in turn, a direct bearing on Earth's climate and weather patterns, a fact that must be obvious from the way EM has tangible effects on our planet, providing us with light, heat, radio and X Ray machines. Naturally-occurring magnetic activity goes on all around us, because Earth, and many other planets and stars, behaves like a giant dynamo. Earth's magnetism has existed as far back as scientists can trace — some three billion years — and our understanding of it has had a highly influential bearing on the subject of plate tectonics. (Tectonics itself, explaining how the continents have drifted into their present positions, has revolutionized our understanding of Earth.) Earth magnetism is the twentieth-century equivalent of the kind of apprehension that nineteenth-century scientists gleaned from the meaning of fossils in rocks; indeed, it is very similar in concept to fossilization.

It is also clear that the source of the field is constantly changing at rates which far exceed usual geological processes. It topples over every few million years, and takes a few thousand years to do so. The only known part of the Earth that can move about so much is the core, and this fits in with what we know about dynamos. It is therefore believed that the molten core of the Earth, some 1,370 km in diameter, conducts electrical flows, in the same way as does running water. Some scientists believe this flow is channelled through the outer rocky layers, or perhaps even by the atmosphere. It is because the Earth rotates that we can perceive, with new sophisticated measuring techniques, eddying motions in the core which can turn it into a *de facto* dynamo producing a field. It is curious to note, however, that the incredibly hot temperatures prevent the core from being a permanent magnet, because beyond a critical point the magnetic properties of a metal are lost.

Where did the electrical field come from? One theory is that it must have formed early on in the embryonic solar system, when dense blobs of iron simply sank to the centre of the planet. An alternative theory, advanced by Professor William McCrea of the Astronomy Centre at Sussex University, says that any large grains of matter, formed at the beginning out of the collapsing dust cloud, would become magnetized by the piercing field of the embryonic

Sun. They could then become attracted to each other to produce a massive core, to attract the remaining grains not yet magnetized to form the outer mantle of the planet. It is the Sun, then, that could have first created Earth's magnetic field. From then on it is maintained in perpetuity by the energy derived from the spinning Earth, like a dynamo.

Another theory suggests that the dynamo is periodically recharged by earthquakes, the evidence of which is found in certain kinds of quartz-bearing rocks which then yield a piezo-electric effect when subject to pressure stresses. A.D. Moore, writing in *Scientific American* of March 1972, said that the positive charge of the atmosphere sets up a downward electric field 'amounting to between 100 and 500 volts per metre on a clear day'. This field can even be heard. According to *Agence France Presse* a 'terrible noise' and 'bluish light' accompanied fiery, glowing meteorites when they hurtled over Madagascar in July 1977. The meteorite which fell on Barwell, Leicestershire, on Christmas Eve 1965 generated radiation which could be sensed by many observers as a buzzing or hissing sound. Many of these small meteors are hit by cosmic rays, which react with the meteor's own atoms, causing it to change form or to become radioactive as it begins to break down.

The evidence of geomagnetism comes from the surface, from particles of rock deep within the crust that were magnetized over aeons of time. So the rock strata provides a permanent record of magnetic fluctuations. The reality of continental drift, too, is the evidence of magnetic reversals in the oceanic lithosphere. Successive layers of basalt may contain a surface layer aligned with the Earth's present magnetic field; other layers would line up differently, zig-zag fashion, showing wandering magnetic poles, even complete reversals.

Magnetism and Life
This is not the place to review the extensive literature on the nature of the human EM system. Suffice it to say that the positive and negative charges of the atoms and molecules in our bodies create a primitive electrical system, and in a sense we become living radio transmitters. Possibly the charged ions in our molecules respond to both the attractive and repulsive forces within the magnetic field. There is hard evidence that biochemical and physiological

functions of both animals and humans are governed by 'bio-clocks', and that these can be affected by changes in the ion count of the atmosphere. Many hitherto mysterious cycles of activity (such as migration patterns) seem to have distinct relationships to seasonal, temperature and solar radiation factors: all these are in a sense extraterrestrial forces.

Scientists have proof that bacteria, birds and fishes have been able to find their direction using the geomagnetic field. Biologists have shown that even the neurones of the brain have magnetic fields. Michel Gauquelin, the French psychologist who for many years has made a name for himself by investigating the claims of astrology, suggests that the similar personality traits of parents and their children are dependent on the state of the magnetic field on the day the child was born, there being a two-and-a-half times greater chance of similarity on a magnetically disturbed day.

Percy Seymour is an astrophysicist at Plymouth Polytechnic, and believes planetary alignments may indeed be responsible for the kind of personal attributes assigned to them by astrologers. For example, he says that the gravity of the moon and the nearer planets cause tides in the Earth's magnetosphere. These produce electric currents which alter the magnetic field in a regularly repeating pattern. Perhaps, he says, there are 'magnetic bays' similar to those that amplify the effects of river estuaries. The rotational and orbital differences of the planets are responsible for distinct frequencies (with overtones) peculiar to each planet, which can make the neural circuits in the brain of human embryos resonate with these frequencies. World leaders, apparently, have their embryos resonating with Jupiter's influence. Critics, however, point out that magnetic tides repeat roughly once every day as Earth turns, but planets move at different speeds, so the repetition is different in each case. Jupiter's 'signal' recurs every 23 hours 56 minutes and 24 seconds, and Saturn's every 23 hours 56 minutes and 12 seconds.

There is much more evidence of the field in action in Earth's biosphere and hydrosphere. Electric storms, and some kinds of dry winds, such as those which howl their way across the Russian steppes to the Alps, are evidence of EM in action. The air is full of positively charged ions arising through the friction of wind, and can knock off electrons from oxygen or water vapour (H_2O). Even the flushing of a toilet can create tiny negative fields. Low-current electrical charges can have pronounced healing properties, too, while waterfalls have long been known to produce beneficial negative ions.

So much for the good news. Physicist John Taylor of London University believes that positively charged particles can accumulate on the surface of living cells, and change their metabolic activity. On Earth some substances with very heavy nuclei become unstable, and undergo radioactive disintegration as their outer electrons begin to escape the hold of the nucleus. They can emit gamma rays capable of passing through a centimetre of aluminium plate. And long-term exposure to EM is known to have an adverse effect on rats.

Modern electronic equipment in homes and offices radiates EM waves resulting in an all-enveloping blanket of radiation, known as 'electronic smog'. EM radiation travels in waves by inducing periodic changes in electric currents in metal aerials, and in turn these currents can find their way into the circuitry of other devices and so cause electrical disturbances. At certain frequencies and strengths they have serious effects, corrupting stored computer data and switching machines on and off. Computer screens and display units also emit radio waves, light photons and other kinds of EM. Hence there is a new fear that computers storing classified information can leak signals that can be picked up by eavesdroppers listening in to extraneous radio waves. At present there is legislation in Britain to control interference from unauthorized radio transmitters and electrical equipment, but not from EM smog. In America scientists have warned of 'electromagnetic pollution'. The White House Office of Telecommunications Policy (OTP) says that this type of waste in a complex industrial society arises from the production of energy itself, which could be the source of unhealthy radiation that could deleteriously affect human behaviour. As a result a new science is emerging which assesses the ability of an electrical device to function with compatibility in an EM-polluted environment.

There is the added risk of harm from events in outer space. Beams of neutrons exist, together with their associated massless neutrinos (said to be able to pass with ease through solid lead *millions* of miles thick!). There are many other sub-atomic particles, all of which are a form of radiation. Beta-rays, which on Earth can penetrate aluminium foil, are fast-moving electrons.

Electrically charged particles which shower down to the surface are said to emanate from the innermost depths of space, probably from the vibrant radio stars called pulsars. And when a large star ends its life in a spectacular super-nova it ejects matter violently to create what are known as cosmic rays — fast moving atomic particles. According to the latest discoveries of physicists at Durham University these supernovae remnants will produce 'rays' with the penetrating power of 10^{13} electron-volts. They invade the Earth with as much radiation as all the other astronomical EM put together, excepting that from the Sun itself. Only gamma rays are more energetic. Durham's Arnold Wolfendale considers that they are probably produced by the proton-boosting effect of a nearby Black Hole. It is now known that beta and gamma rays can cause mutations, and speed up the rate of evolutionary change. They possess far more energy even than those created by nuclear reactions. As you begin to read this sentence, for example, some 50 cosmic rays will already have passed through your head. They have been known individually to penetrate three feet below the surface of the Earth.

Astronauts have to face constant danger from cosmic radiation. A team of three US radiation scientists reported in *Nature* magazine in February 1988, that outside the Earth's protective magnetosphere cosmic rays would give astronauts doses of up to 50 rem per year, 100 times more than the radiation dose that scientists think would be within the safe margin to avoid getting cancer. Unpredictable solar flares could suddenly increase the lethal dose of up to 1,000 rems *per day*. The team suggest that in any foreseeable manned-voyage to Mars the habitable areas of spacecraft be shielded with at least 7.5 cm of aluminium or its equivalent.

What solar radiation and magnetic field reversals can do to Earth and what the climatic effects might be we will leave to a later chapter. In the meantime, having underlined the potential dangers arising from the forces of energy in the Cosmos, we ought to take a closer look at our Sun and its internal dynamics. In recent years rumours have abounded: the Sun is fading fast, it is wobbling too much, it may not be a 'normal' star, and its recent erratic behaviour may explain the extremes of weather the globe has been suffering from in recent years. What is the truth behind these rumours?

Chapter 2:
HOW WELL IS THE SUN?

In the freezing cold month of February 1986 an editorial in one of Britain's daily tabloids started with the alarming words: 'The tired old Sun is going steadily downhill, and that's official'. The Sun was apparently losing lustre at .02 per cent a year, and had even started to wobble. It could, said the tabloid, trigger an Ice Age in just 500 years, which could be so desolate and inhospitable that 'all life would be extinguished'.

This item, fortunately, was an oversimplified interpretation of an article that appeared in *Science* magazine of that month. For regardless of what satellite readings of the Sun's brightness then showed, the Sun would have been fading too fast for it to be part of any long-term change. In the meantime the Sun will continue to fluctuate and wobble, radiating as it has done for aeons on a variety of spectra. So while one form of energy might be diminishing, other forms might actually be increasing, as was hinted at by another report of the latest measurements of the Sun from the Pasadena Jet Propulsion Laboratory, which said the Sun had expanded by about 100 kilometres.

One of the most alarming suggestions of recent years is that the Sun may not be a 'normal' star. The kind of regular variations in radiation it undergoes is similar to those stars known as Cepheids, which fluctuate periodically in a way that depends on their mass and luminosity. Time-worn assumptions about the Sun are already being modified in the light of recent discoveries, and many scientists have now abandoned the phrase 'solar constant' in favour of 'solar parameter', since the Sun's radiation seems to be anything but constant.

True, the Sun does go through disturbing cyclical phases, sometimes showing signs of chronic instability, which scientists reassure us is nothing to worry about — yet. The instability may be perfectly natural. Some solar physicists believe the nuclear furnace of the Sun may not 'burn' constantly but fluctuate, like a thermostat, in response to conditions in the surrounding galaxy.

Scientists certainly know more about the Sun than they did even ten years ago, because of the improvement in high resolution observations made from the ground and from spacecraft launched by many European teams, backed up by excellent British and American theoretical deductions.

One well known US climatologist believes the Sun virtually 'breathes' in and out every 76 years. Ronald Gilliland of the National Centre for Atmospheric Research, Boulder, checked back into the historical records of the Sun's diameter. Surprisingly, data about the Sun's size goes back to the early eighteenth century — time enough to study the duration of eclipses of the Sun and Moon, and even the passage of the planet Mercury across the face of the Sun. From the data available Gilliland could calculate that the Sun's diameter oscillated on an unmistakeable 76-year cycle, between being slightly bigger and slightly smaller. Gilliland is unclear as to the cause of these cycles, and makes an unhelpfully obvious point: there is a spectrum of repetitive phenomena occurring deep inside the Sun. What is interesting, though, is Gilliland's estimate of the size of the Sun: it was at its biggest in 1911, and at its smallest in 1949. It would have been, he said, at its biggest again in 1987. Gilliland, in fact, believes that the cyclical Sun may be interfering with our understanding of anthropogenic warming influences. The Greenhouse Effect, he says, may be counterbalanced by periods of shrinkage in the past, and by implication the present global warming may be the product of the maximum size Sun plus the Greenhouse Effect.

In one sense, of course, the Sun will always be losing power. Helmholtz, in 1854, discovered the law of conservation of energy, which means that whatever the power source of the Sun, the Sun must eventually die. But Helmholtz believed that the solar energy was provided by gravity itself, so that as the Sun contracted under the pull of its own mass the energy so delivered would increase with greater and greater intensity. This is not the way the Sun's heat is actually generated, but it was a remarkably visionary explanation considering the state of physics at the time, and foreshadowed Einstein's mass-energy theorems. In any event, Helmholtz was right in the sense that the Sun is a giant dynamo converting mechanical energy into EM energy. In this case it is the ceaseless whirl of the charged particles that act as the dynamo.

The problem with this explanation, as later nineteenth-century scientists found out, was that the shrinking rate of the Sun would be too fast — collapsing under gravitational contraction from a diameter of 186 million miles some 25 million years ago to its present size. Helmholtz's theory implied that the Sun 500 million years ago would have filled the sky. And yet photosynthesis was going on 700 million years ago, and the Earth was not being scorched to a cinder. Further calculations show it would continue to shrink to the size of a marble in about 250,000 years from now. However, in 1929 the American astronomer Henry Norris Russel proved that the Sun was predominantly hydrogen, and that this hydrogen existed in a plasma state. We now know that the source of the Sun's energy, a source that means it will last much, much longer than 250,000 years, is the continual *fusion* of elements. For when the Sun contracts, Helmholtz style, to its present size, the core grows hot enough to initiate nuclear reactions.

The simple explanation as to why the Sun lasts much longer than was previously thought, is that energy in an atomic nucleus is millions of times greater than that provided by purely chemical processes. Indeed, the nucleus forces are very much more powerful than that holding the electrons in orbit (i.e. the strong nuclear force compared with the electromagnetic force). The Sun's heat is created by nuclear explosions going on at its core. The electrons from the hydrogen atoms are stripped off, leaving behind individual protons. These protons (even though they are electrically positive and normally repel each other) soon fuse, or blend, together because of the overwhelming kinetic energy of the hot plasma, reaching 15 million degrees at the Sun's centre. Through an intricate process, some of the protons and neutrons become the nucleus of deuterium (a heavier form of hydrogen), and then further fusion adds more protons, to turn the deuterium into helium. Mass is converted into fearsome heat in the process, and using Einstein's famous equation $E = mc^2$ we can calculate that the Sun is losing 563 million tonnes a second. An estimated 89 quadrillion (89×10^{24}) tonnes of hydrogen have so far been transformed into helium, and the sun has exhausted one-third of its nuclear fuel in the process, and will continue like this for an amazing 5 billion more years.

Even when the sun approaches the end of its long life other compensating factors would come into play: for instance there is the chemical composition of Earth's atmosphere that could slow down the rate of heat loss. And we must remember that when the sun shrinks, its gravitational force is naturally reduced. This itself could mean that Earth will cool as its orbit becomes lengthened; as it rotates around the Sun further out in space.

To compensate, however, the Sun — as it undergoes its death throes — would turn into a Red Giant and actually become warmer. This is because a Red Giant star has a core which has grown richer and denser in helium, hence the gravitational field in the core contracts and forces up the internal temperature by something like a factor of 1,600. The helium nuclei combine into ever-more complex elements, such as carbon, oxygen and silicon — the building blocks of all planetary matter. The star expands in size as a consequence, but the radiated energy is spread over a wider area, and becomes red-hot instead of white-hot. Finally, when the Sun's nuclear fuel is spent, its total weight will be compressed into a superdense sphere about the size of our own Moon — a deathly cold Black Dwarf.

All this, of course, will take place far into the future, and it hardly seems possible that humanity, as a species, will still be around to worry about it. (The reasons the sun's nuclear fusion process takes so long to consume mass is because of its colossal size. It has a mass of about 335,000 times that of the Earth, although its diameter is only 100 times greater. But its volume is so large that one million Earths could easily fit inside it. The whole of the rest of the solar system would comfortably fit into less than one per cent of the Sun.)

Space Dust and the Sun
Periodicity is rife in the universe, and according to Bill McCrea of Sussex University herein lies a plausible linking factor between the behaviour of the Sun and the ice ages. McCrea assumed from the geological evidence that great ice ages occurred every 250 million years or so — coincidentally or not, the precise time it takes the solar system to circulate once around the Milky Way. The Sun was formed five billion years ago, and has circulated around the galaxy ever since. In so doing it has crossed one of the spiral arms, the Orion Arm, some fifty times in its life. About a million years ago, however, it entered a dark 'compression lane' situated along the edge of the spiral arms, and exited it at about the time the

Quaternary ice age came to an end. This fact has prompted imaginative new theories about how the Sun might 'warm up'. McCrea theorises that every time the solar system passes through this very dense part of the galaxy, the volume of space between the Sun and the Earth would become about 100 million times more dense than usual. The Sun would begin to absorb dust into itself, plus some of the additional hydrogen molecules in the dust lane. Then the surface of the Sun would warm up.

Not all scientists agree with McCrae's theory, however. An important criticism highlights the tiny distance (by intergalactic standards) between the Earth and the Sun. This would be too cramped even for a 100 million-fold increase in its average density to have much effect on the Sun's strength. The spiral arm line would produce a total mass of scattered particles amounting to only 10^{15} tons, about one six-millionth of the total mass of the Earth — too small to interfere with the warmth that we get from the Sun. However, the Sun may meet up with additional clouds of interstellar matter drifting randomly through the galaxy. Francesco Paresce and Stuart Bowyer of the Space Telescope Science Institute at Baltimore, have suggested that the Sun is currently immersed in an intergalactic cloud of hydrogen and helium coming from the direction of the constellation Centaurus. This could have a cooling effect on Earth, as we will discuss in the next chapter.

The Missing Neutrinos
At the moment, however, all discussion of a warming or cooling Sun is somewhat academic. The difficulty in deciding whether or not the Sun is losing power lies solely in our ability to mathematically compute accurately the energy produced at its core. One problem with this arises from its plasma state.

The Sun is actually a second generation star built up from cosmic dust grains and the helium particles left over from other stars that blew themselves to pieces aeons ago. So it already has a lot of helium.

The late eminent astronomer Ernst J. Opik believed that the creation of these heavier substances explained the Ice Ages. Within the Sun they act as a barrier to the energy radiating at the core. The core then becomes enlarged and uses up much more energy in the process. As a result less radiation reaches Earth, causing periodic Ice Ages. This theory, however, has been challenged by astronomer Roger K. Ullrich of the University of California at Los Angeles, who believes that the metallic elements could not move rapidly enough to reach the core in its own lifetime. But helium-2, a heavier form of helium, could do the trick instead.

Indeed, over the years a lot of theories about the Sun have been challenged. How do we actually *know* what goes on inside the Sun? The problem with science is the way theories often need to be backed up with mathematical models. This of course provides 'proof' of propositions, and with subjects like astronomy and cosmology, when one is dealing with events taking place millions of miles away in spacetime, there is no other way to formulate theories. This is all very well until the theory fails to account for disturbing anomalies and discrepancies that cannot be examined empirically. For example, according to the maths, it take anywhere from one to 10 million years for photons (light particles) to fight their way through to the surface of the Sun, since they are constantly being absorbed and re-emitted by other matter. So while we wait for the particles to bombard us, the theory of solar fusion remains just that — a theory. And according to something known as the standard solar model the Sun should be emitting vast quantities of neutrinos.

Neutrinos are supposed to be the product of nuclear reactions going on in the Sun. Whereas the Sun's heat and light take one million years to fight their way to the surface, neutrinos take only a couple of seconds. They are said to be the product of beta decay, a process that breaks up atomic particles and sends out speeding electrons. But in a sense neutrinos are the inventions of physicists (in this case an Italian called Enrico Fermi), like some other sub-atomic particles discovered in the 1970s and 1980s, to 'balance the books', to provide for symmetry. The problem was that the amount of energy lost from the nucleus was more than could be accounted for by the escaping electrons, so each electron was assumed to be accompanied by a neutrino (or 'little neutral one'). But being massless it was virtually indetectable. Some claim that neutrinos change their identity while speeding away from the Sun's photosphere or from its core. As Frank Close of the University of Tennessee says, there might be some doubt in regard to refraction or spin as they pass through matter of varying densities.

Many exotic experiments have been done to prove the existence of these mysterious particles, including, surprisingly, trying to detect them passing through a tank of cleaning fluid! They are said to be so insubstantial — less than a millionth the mass of a proton — that they could well be no more than a figurative abstraction based upon what has been discovered going on in particle accelerators. Because they virtually never interact with other particles, they can shoot out of the Sun to actually pass through the Earth.

Even so, a neutrino can occasionally be caught if enough complicated obstacles are placed in its path. Neutrinos are extremely rare; there are probably no more than 500 in every cubic centimetre of space. So they are overwhelmed by the 'noise' of other particles — cosmic rays from space, plus the protons, alpha and other particles emanating from the Sun. But there are some very rare substances, like one called gallium, that can respond to neutrinos. And in some nuclear reactions on Earth they can be made to change into electrons. But tanks of cleaning fluid containing billions of chlorine atoms deep in mines in the bowels of the Earth seem to be the best bet, and the most famous experiment to 'catch' neutrinos in this way was performed by Raymond Davis of the Brookhaven National Laboratories in a rock cavern in South Dakota.

Yet still there are not enough neutrinos to tie in with the current theoretical solar model. Scientists are baffled and not a little disturbed. If there are no neutrinos, or if there were neutrinos once, but no longer, does this mean that the Sun is dying? Or does it confirm our fears that the Sun is not an ordinary star after all, or will not evolve into its Red Giant phase? If the rate of neutrino production is supposed to depend on the Sun's heat, varying by a multiple of ten, does this mean that the Sun is less hot than we think? (or rather, have we worked out correctly *where*, in regard to the different parts of the Sun, the heat is coming from?)

More than likely the problem resides within Man himself and his equipment, i.e. either with his mathematical constructs or his overly-sophisticated computer technology. Detecting solar neutrinos has now been raised to the status of a philosophical problem; indeed sociologists now write books on the subject. The difficulty lies not only in identifying Nature's message, but rendering it into comprehensible maths and, a much more difficult task, into plain language. Idealized accounts are clearly not easy to reconcile with the complexities of research, and it would help if only one energy spectrum was used. Frank Close, for example, recently pointed out that there are 'low energy' and 'high energy' neutrinos, and Japanese scientists investigating them coming from exploding supernova counted far more than their American counterparts.

However, a new theory that attempts to overcome this awkward anomaly of the missing neutrinos centres around the invention of a new particle — the weakly interacting massive particle, or WIMP. Indeed, because other matter cannot restrain them — because they interact 'weakly' — they can theoretically penetrate a solid lead wall more than 100 million miles thick at the speed of light! (which, incidentally, is a property also ascribed to neutrinos). And, like neutrinos, they are extremely scarce in the universe; American astrophysicists believe there is only one Wimp for every hundred billion particles inside the Sun. First proposed in 1985 by John Foulkner of the University of California, Santa Cruz, and Ronald Gilliland of the National Centre for Atmospheric Research, they are said to collide alternatively with hot and cool protons, exchanging heat energy outwards from the inner core to the outer core. This theory thus satisfactorily explains how the Sun's heat gets down to Earth, and why there are less neutrinos than there 'should' be.

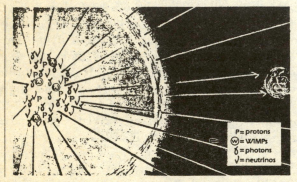

If WIMPS inhabit the sun's centre, they may carry heat away from the inner core and thereby limit the number of neutrinos created — which would agree with observations.

Source: *Science Digest*, March 1986

The Magnetic Sunspot Clue

Although the internal workings of the Sun may occasionally be subject to debate and speculation, knowledge about what goes on on the surface is much more certain. Much of what we understand about the history of the Sun's behaviour is derived from direct observation, dating back to little more than 500 years. Through a telescope the Sun displays evolving patterns of 'flames', hundreds of miles high. The Sun shoots out tiny bits of itself into space, namely hydrogen nuclei (protons) and negatively charged elementary particles (electrons). But they can be flung out in three different ways.

First there are the flares which often combine with much brighter *faculae* (torches). Flares are extremely hot; one theory is that they are contained for a time by the magnetic field of the Sun, but at times break free. Occasionally a massive surge of blazing gas (the spicules) is seen travelling at a rate of 660 miles per second or so. Sometimes nearly 10 flares a day can be observed. They can reach temperatures markedly higher than that at the surface — some have registered 20,000°F or more. Many can reach up to 100,000 miles beyond the Sun's photosphere, a 290-mile thick layer of gas with a temperature range varying from 18,000°F nearest the surface to about 7,560°F at the top. (The Earth also has a record of solar flares. Tree rings can be dated according to the amount of carbon-14 in them, which is produced when a neutron knocks a proton out of a nitrogen-14 nuclei. Hence solar flares are thought to create carbon-14 in the atmosphere, a theory which has enabled Paul Damon of the University of Arizona to record most major twentieth-century flares.)

Then there is the solar 'wind', only discovered in the 1940s, which is a continuous outpouring of high-energy particles at temperatures of more than a million degrees.

But it is the curious dark blotches on the Sun, known as Sunspots, that stimulate the most scientific interest. They have been suspected for a long time of having some connection with the Sun's variable output. The size of the spots themselves changes quite frequently; although many are only a few hundred miles across, some can range up to 10,000 miles in diameter.

There is an unmistakeable connection between sunspots and flares — as the leading spot in each group first becomes visible, so do the flares. They seem to well up from a mighty expansion of gases emanating from the core of the Sun, bubbling up from a hot, molten river. It is this expansion — using the same principle as that of the domestic refrigerator — which causes the temperature to drop. The darkness of the sunspot, however, is illusory. It simply appears dark against a bright, glowing surface, albeit some 2,000°C cooler than the rest of the Sun's surface, and the reason for the darkening on the surface is largely due to the curling magnetic forces whipping up the electrically charged particles, with massive flares being flung outwards. Ironically William Herschel, the most eminent astronomer of his generation, declared in 1795 that the sunspots were holes in a blazing outer corona, through which a cold gaseous interior — rather like Jupiter today — could be seen.

Spots appear in groups, often dominated by a prominent pair which seem to grow, cell-like, until they absorb the smaller spots. Some spots can last months, others only a few hours. They emerge alongside the magnetic lines of force with their own fields at least 1,000 times stronger than both the field of the Earth and the rest of the Sun. More importantly, they predominate over an eleven-year cycle, a cycle which occurs because of the poles flipping from north to south. The strength of the field also varies considerably; when it is 'active' it is the field that becomes visible as sunspots. Each pair of spots is magnetically polarized, with each one having a north or south polarity. Whatever the north polar or south polar arrangement, the situation in the opposite hemisphere is curiously reversed. In other words, if a right-hand spot in the northern hemisphere is s-polar, then a left spot in the south will also be s-polar. Not only that, but after the eleven-year cycle has ended, all the matching polarities are reversed for the next eleven-year cycle.

The Sun's magnetism has long been suspected of having a wide-ranging impact on Earth, affecting, in particular, its biosphere and climate. This impact became better understood in the early postwar years after RCA Communications Inc. of America set up a small observatory and laboratory on the roof of its main building in central Manhattan in 1946. RCA had known for some twenty years that the Sun's behaviour, plus magnetic storms, could interfere with radio communication, so the company had a vested interest in learning more about the phenomenon. John Henry Nelson, an engineer of long-standing with RCA, was given the task of predicting good 'radio weather' so that RCA staff could be forewarned of reception difficulties.

Nelson first of all concentrated on sunspots, since they had been observed for centuries to come and go in cycles, and perhaps, following the discovery in 1908 by American astronomer George Ellery Hale of solar magnetic forces, had something to do with the periodicity of radio interference patterns. Soon Nelson noted that it was the occasional 'maverick' spot that could wreak havoc with the airwaves, and he began to predict radio weather of an accuracy of up to 70 per cent. But Nelson made other important discoveries. It was becoming clear that the planets had something to do with sunspots, and that they were closely connected with EM, confirming the suspicions of the erstwhile Sir John Herschel in 1833. Nelson concluded that the Sun was, in fact, little more than a giant armature hanging in space, with the planets acting like celestial magnets. He could see the similarity between an electric generator and its magnets, and the Sun and its rotating family of planets. However, because the planetary 'magnets' are moving, this proved to Nelson that the EM energy of the solar system fluctuated widely. Indeed, his evidence seemed conclusive. On 12 April 1951, RCA issued a press release announcing a direct relationship between magnetic storms on Earth and the behaviour of the Sun and planetary positions.

Return to Jupiter
We can see now how the Jupiter Effect, and other gravitational theories, could actually work. We have discussed in the previous chapter how objects in space have perceptible and measurable influences on each other. Edward Harrison, of the University of Massachussetts, believes the Sun may have a 'dark companion', as yet undetected. We have now introduced, ahead of time, the 'Nemesis' death-star theory, of which we will have more to say later in this book. This theory says the companion star sets in motion a great army of comets winging their way towards Earth's surface. An alternative theory, however, says that the twin star, instead, would very likely have a comet-like orbit, taking about 10,000 years to do a single revolution. Hence, at about 5,000-year intervals, it could have both an EM and a gravitational influence on the Sun, to cause it to lose power, and in turn bring about widespread distortions to Earth's climate.

The Jupiter Effect, in the meantime, dictates that the planets can produce minute bulges on the Sun, both front and back, in much the same way that the Sun and Moon affect the tides of the planet. The motion of the Sun around the centre of the solar system is evidence enough of planetary influences. Now, physicists Rhodes Fairbridge and Jim Shirley, writing in the December 1987 issue of *Solar Physics*, have described what goes on using a dumb-bell analogy, with the weighted ends undergoing complex oscillations. The Sun, in fact, is being tugged across a distance greater than its own diameter by Saturn and Jupiter. In the meantime, as the combined planets circle the Sun, (or, rather, the huge gravitational vortex at the very centre of the solar system), they produce individually a tiny pair of bulges on the Sun's photosphere as it

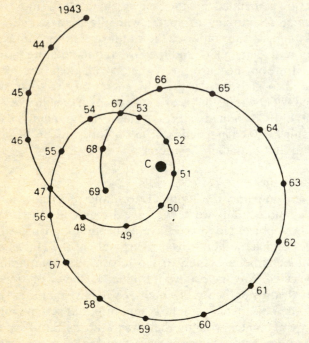

The Sun is not a fixed central feature of the solar system, but moves slowly around the system's centre of gravity (C). This diagram shows the spiralling orbit of the Sun itself from the years 1943–69.

Source: *Judgement of Jupiter*, Richard A. Tilms, Book Club Assocs, 1980

performs its own small eliptical orbit at the centre of the vortex, one on each side of the Sun facing towards the planet. The 'stirring' effect pushes the centre of mass around the central plane, with the Sun swinging around in a loop-like orbit, and it is this that affects its innards. As insignificant as any solar blemishes arising from the curious configuration may be, the odds are highly likely that certain overlapping positions of the planets must cause a combination of bulges, or at least affect the *timing* of the bulges. Physicists at Denver University have successfully related proton bursts on the Sun with planetary line-ups that would add up the solar bulges to make one huge flare.

Jupiter, then, being the biggest planet (especially when it is in conjunction with Saturn) must have the greatest effect. Jupiter is now clearly linked to the eleven-year sunspot rhythm. And if the influence of Earth is added in, the tidal forces are further strengthened. K.D. Wood of Colorado University found that when Earth, Venus and Jupiter are in conjunction or opposition they can increase solar tides up to 50 per cent or more. Ernst Opik

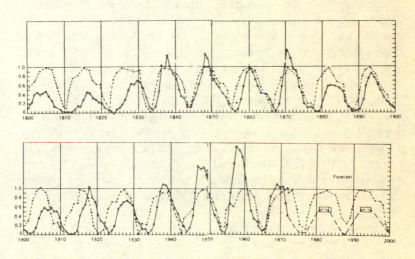

K.D. Wood's evidence for a link between sunspot activity and planetary alignments, since 1800AD. The broken line shows how the sunspot varies rhythmically, and the solid line shows the tidal influence on the Sun.

Adapted from John Gribbin's *Forecasts, Famines and Freezes*, (1976)

calculated the gravitational effects of these three could raise a tidal flow on the Sun that 'is not at all negligible'. Professor Wood calculated the tidal pull on the Sun from Venus and Earth, combined with that of Jupiter, probably caused the giant solar flare that was observed from many observatories across the globe in 1972. Indeed, the chance of the 1972 flare being entirely spontaneous was calculated to be one in 2,000.

We must bear in mind the fact that the Sun is plasma. If it were more like a solid planet its magnetism could be trapped largely below the surface. But the outside slips and slides round at a slower rate than the inner core, forcing a pair of torus-shaped magnetic fields to spin unevenly around beneath the photosphere like a giant lopsided wheel. Convection currents then shift the fields up and down as the Sun rotates.

Hence a crucial combination of bulges could easily speed up the outer flow of the Sun, thus precipitating unusually large flares. True, the figures do not match up as precisely as many proponents would like. But one important clue is the sunspot cycle itself — the way the sunspots grow in number from a 'minima' to a 'maxima'. This is why the magnetic field is created — the split-up of the charged particles in a swirling motion creates the field which alternates on a twenty-two year cycle. During the twenty-two year magnetic cycle sunspot numbers peak, while the magnetic polarity of sunspots reverses to the opposite sign and back again. W.H. Portig of the National Engineering Science Company believed he could prove that Jupiter and Saturn's synodic configurations were linked to spot formation, since after extracting their synods from the sunspot cycle, he found an unexplained gap of 11.1 years. And as early as 1900 E.W. Brown, of Yale University, once plotted a curve showing Jupiter and Saturn in conjunction and opposition, and also 'square' to each other and at 90 degrees apart, going back 300 years.

Computers can now trace these changing patterns back over the centuries by working out the tidal pull of the two big planets. Astronomical and carbon-14 tree-ring evidence suggests the Saturn and Jupiter cycles are 179 years long. Of course, the fact that the eleven-year sunspot cycle is the same as Jupiter's orbital period could simply be coincidental. But if other conjunctions involving Jupiter, Saturn and others are taken into account a distinct statistical relationship (of which we will have more to say later in this book) occurs. Of nineteen such synods noted since 1600, the earliest date at which weather records were systematically written down, eight have occurred during the northern hemispheric summer.

Here, then, we have an important clue to the mystery of why the Sun occasionally appears to be going downhill. It is the planets that are implicated in the sunspot periodicities. It could then be the planets themselves which cause Ice Ages in ways not entirely understood. But climatic change may also be connected with the Sun's EM, and the way it interacts with earth's own EM field.

It is time, now, having briefly looked at how the Sun functions, to move on to see how all the turbulence and movements we have been describing can produce changes in Earth's climate.

Chapter 3:
EARTH'S COSMIC WEATHER CYCLES

How do scientists push forward the frontiers of human knowledge? One way is to attribute causal connections to curious and inexplicable events, if other discoveries about nature hint that they have a lot in common with each other. This commonality, in a sense, implicates the various causal factors under review. In police terms the evidence, although circumstantial, is nevertheless incriminating. For example, during the past 400,000 years there have been important correlations that have matched up with periods of global cooling. Most of the 'Milankovitch cycles', of which more later, are derived from similar cosmic correlations, having do to with warming and cooling episodes relating to orbital configurations. It is also known that the power to turn the weather machine is equal to that released by earthquakes. In addition the amount of energy needed to slow the Earth's rate of spin even by a minuscule number of milliseconds is also the same as that released by earthquakes. Is this just a coincidence? Is it also a coincidence that the disastrous quake at Agadir, Morocco, in 1960 was preceded by a massive solar flare the year before? Could such events be the demonstration of a hitherto unknown physical law, or set of laws?

The evidence for cosmic inputs grows firmer every year. It does seem that some prolonged bouts of inclement weather have unexplained repetitive characteristics that cannot be accounted for by analyzing normal weather dynamics. In a *Nature* article, Californian geologists R.L. Rosenberg and P.J. Coleman once suggested that rainfall patterns in Los Angeles dating back to the year 1900 were curiously concentrated in the first half of a twenty-seven-day period. Not only that, but they gradually shifted the phase of the cycle from year to year. They believed this might be due to cyclical phases in the jet stream that snakes some 40,000 ft high in the atmosphere. The streams themselves, suggest the authors, also have a periodicity that correlates vaguely with solar and lunar phases. Furthermore in October 1986 *New Scientist* reported that an Australian climatologist, Bob Vines, had published evidence on the influence of the twenty-two-year sunspot cycle on repeating rainfall patterns in India. Vine's study used a detailed statistical technique called Maximum Entropy Analysis (MESA), and his findings in India matched up with similar cycles he discovered in Europe, South Africa, South America and Australasia. In the same issue *New Scientist* reported that the Norwegian Geotechnical Insitute had discovered that avalanches in Norway over the past 130 years displayed a periodicity of 12–13 years long, and which had something to do with severe winters which also follow a similar recurring pattern.

True, the flowing nature of weather patterns, and their sometimes idiosyncratic and violent impact upon human society, tends to reinforce a sense of imprecision and irregularity. Bill Burroughs, for example, a physicist and former atmospheric scientist, reminds us that the weather 'machine', in maintaining an overall balance, has inputs only vaguely related to outputs. As a result, weather events can fluctuate on every conceivable time-scale, he wrote, and be notched up with umpteen different solar and planetary alignments.

Precise long-range weather forecasting, for this reason, is difficult, if not impossible, because the connection between cause and effect is very tenuous. Any tiny initial putative cause can have a knock-on effect that is literally too small to measure. *In principle* climate and weather changes can be determined by the beating of the wings of a parakeet in the forests of Amazonia, and mini tides around the world's coastlines can be determined by someone dipping their big toe into the sea at Eastbourne. Mathematicians actually use the word chaotic to describe such phenomena. They know that the smallest uncertainty in initial conditions can be amplified to produce chaotic, unpredictable results. Instead of a perfectly ordered universe running along predetermined paths, mathematics shows that we can never have exact knowledge. Sadly, in regard to the theme of this book, chaotic behaviour was first discovered in maths models of weather systems. But any system, even one following mathematical rules, can reveal randomness. A system, like the weather, that is dominated by randomness is even less amenable to mathematical treatment.

And yet scientists, with their imaginations stirred, can see theoretical advantages in chaos, and are using it to study fusion reactors and the orbit of planetary moons. At the Non-Linear Systems Laboratory at Warwick University they are hoping to bring order to chaos, where a technique known as fractal geometry can convert irregular patterns into understandable rhythms of self-similarity.

We must not be too harsh in our judgement. The study of climate is a science, whether precise predictions can be made or not. Weather and climate, as we have seen, follow cycles and display periodicities, and this is to be expected. As all matter in the cosmos and the solar system functions acording to a distinct periodicity it follows that it would not be unusual if periodicity is detected in phenomena here on Earth.

Of course the way the hydrosphere — the atmosphere and Earth's surface water — responds to lunar tugs has been understood for centuries. Our discussion in Chapter 1 of gravitational and EM forces now begins to have more relevance. Firstly, it is worth bearing in mind that the Moon's pull on our planet is 106 times that of Jupiter at its closest. The surface of the Earth facing the Moon is nearly 4,000 times nearer the Moon than the Earth's centre (i.e. roughly half the Earth's thickness). So the lunar pull on this surface is 7 per cent greater than on the far surface. The Moon moves in its orbit in the same direction the Earth spins, and the great bulge of heaped-up water on the Moon-facing side moves with it. And as the Earth turns in its orbit the continents pass through the bulge.

The Sun, too, plays a small role, and gives a mini-tide when the Earth's fluid envelope is subject to the Sun's gravitational pull. The Sun's influence is only .46 times that of the Moon because, although it is 27 million times as massive as the Moon, it is 390 times as far away, since the tidal affect decreases by the cube of its distance.

Naturally, then, the danger arises twice a month when the Earth, Moon and Sun are in very nearly a straight line. The tides then generated are called 'spring' or 'syzygy' tides, and can run up to six feet higher than normal. In the past this has brought about devastating floods around the world's coastlines and along river estuaries. Sometimes the water gets boxed in by narrowing headlands, so that the onrushing tide becomes a tidal bore. These can often, as in the case of the Tsientse Kiang river in China, reach up to 25 feet. (One of the most notorious tidal floods in recent history occurred on both sides of the narrow Flemish Bight in 1953, taking the lives of 2,300 people.) And this is not all. The tidal effects can be greatly enhanced every 18.6 years when two other factors come into play — perigee, when the Moon is closest to the Earth, and perihelion, when Earth is nearest the Sun. Bob Currie of the State University of New York has also found lunar influences on rainfall patterns (due, he says, to 'lunar nodal tides' in the atmosphere). It is known that the inclination of the lunar orbit is five degrees to the ecliptic; the ecliptic is defined as the apparent motion of the Sun across the stars. The inclination of the lunar orbit as seen from the Earth varies between 18.5 and 28.5 degrees, enough to produce atmospheric 'tides'. Indeed, there was much alarm in the United States in the first week of January 1987, when tidal waves and blizzards swept across the nation, destroying whole streets in eastern coastal cities during this rare astronomical occurrence. Tides were higher in Britain, too, but the country was spared flooding because of a ridge of high atmospheric pressure that moved down southwards, bringing cold but settled weather.

Earth's Electrical Cooling System

The Sun's importance in determining weather events is not just when it plays a part in an Earth–Sun–Moon alignment. It is of vital importance in setting the entire 'weather machine' in motion. However, what is not so obvious is the crucial role solar fluctuations play in bringing about long- and short-term changes in the world's weather and, ultimately, upon the climate.

One thing we have learned about our study of the Sun is that solar fluctuations themselves do not always imply variations in infrared energy; they do not mean that more heat will reach Earth. The Sun may indeed get 'cooler' when sunspots disappear, but such an event would hardly be noticeable at the Earth's surface. Indeed, given the natural variations in seasonal temperatures and any possible trends towards a global warming brought about by Man's own activities, it would be impossible to detect. There are, of course, occasions when the Sun does instead 'warm up', in the sense that we understand the term, and such a warming will naturally have repercussions here on Earth. A detailed discussion of the way solar heat itself can actually — and paradoxically — bring on an Ice Age will appear in the next chapter.

There is, however, something more subtle and intangible concerning Earth's relationship to the Sun that we must dispose of first. Let us for the moment turn our attention to Earth's atmosphere. Our windy envelope, in simple terms, can be divided into the outer layer — the stratosphere — and the layer closest to the surface — the troposphere. The interaction of the atmosphere with the Sun is regulated by the altitude of the various layers of air. The outer layers reflect back a little of the radiation. From there the heat rebounds back to warm the lower atmosphere, but it is partly trapped by the carbon dioxide. This creates the well-known Greenhouse Effect, which keeps Earth warmer than it would be without its atmosphere.

More importantly, when the Sun's rays strike the stratosphere a photo-chemical reaction takes place. The oxygen molecules are bunched together in threes, instead of twos further down in the troposphere (03 instead of 02). But is it just the Sun's thermal radiation that causes the photochemical reaction? The answer is a little complicated. Heat is one form of invisible radiation. And, along with the other wavelengths emanating from the Sun at the short and long ends of the spectrum, heat is absorbed all the way down through the gaseous layers to the surface. But one thing we have learned from Chapter 2 is that the Sun is not 'constant', and that it is prone to flaring and sunspotting.

The atomic nuclei that the Sun ejects violently into space are forms of 'cosmic rays', although of the 'softer' variety. These solar rays are also referred to as the 'solar wind', and apply to other stars as well; the hotter and larger the star the more energetic is the wind that is thrown out. These cosmic rays penetrate quite far into the atmosphere, and often reach the Earth's surface. Numerous secondary electrons are produced, called electron showers, from the ionising collisions with atmospheric molecules. The secondary electrons in turn often ionize molecules to form negative ions. The result is that one cosmic particle could create as many as one billion ion pairs. Furthermore, a potentially grave hazard to all biological life on planetary systems is when stellar objects spray cosmic rays out into space more violently than usual. One of the most violent events is a supernova burst, when a star ends its life in brilliantly explosive death throes.

Another potentially hazardous event to life on Earth is when the Sun itself flares up more violently than usual. When this occurs there is a knock-on effect in the atmosphere, similar to ionisation, but instead of atoms being torn apart the larger molecules of gas are split into atoms. The pulverized atoms are suddenly prevented from assuming their threesome state, and are turned into ordinary oxygen. The ozone level — consisting of oxygen in the tri-atomic state — is dramatically reduced in the process. Hence, if a supernova was to explode at a distance of no more than thirty-two light years, giving off far more cosmic rays than the Sun does, the amount of oxides of nitrogen generated would be enough to cataclytically destroy ozone. The precipitating agent would be the million ergs per square centimetre of ionizing radiation that would race away from the supernova.

Herein lies the clue to climatic change. According to G.C. Reid and J.R. McAfee, writing in *Nature* magazine, depletion of the ozone layer brings about cooler global temperatures because of the way the delicate balance of chemicals in the atmosphere has been destroyed.

Nitrogen and oxygen make up 99 per cent of the gases in the atmosphere. But the shrunken ozone layer has the effect of causing the gases to react together to form nitric oxide (NO). Then, with the adition of another oxygen atom it becomes nitrogen dioxide (NO_2) (alternatively, an extra nitrogen particle would turn the gas into nitrous oxide, N_2O). Hence a solar flare, by first interfering with the molecular nature of the ozone layer, can generate as many oxides of nitrogen (collectively known as NOx) as a 50-megaton explosion.

The reality of what NOx can do to the atmosphere was demonstrated by Russian scientists. Studies done by Kondratyev and Niklosky proved that nuclear testing in the 1950s and 1960s created some 3,000 tonnes of nitric oxide for each megaton of TNT exploded at the surface. A nuclear blast yielding some 340 megatons was released in the run-up to the test ban treaty of 1963, and at the time there were massive amounts of oxides of nitrogen polluting the atmosphere. Was it simply a coincidence that skies in the northern hemisphere, between 1958 and 1964, seemed to be much more overcast than usual? In any event a supernova burst, and to a lesser extent an episode of solar turbulence, could easily create a colder world as high altitude cirrus clouds freeze into particles and reflect back much of the Sun's heat. A great many creatures, of course, would not necessarily come to harm (aquatic creatures, like fish, are likely to survive the best: the thermal inertia of the oceans — taking up to a century to produce a temperature change of just a few degrees — would protect sea creatures from any catastrophic changes of climate on land).

However the supernova threat is constantly with us. At least five historically confirmed supernovae have occurred in the last 1,000 years in that part of the galaxy nearest to us. According to Arnold Wolfendale and his team of astronomers at Durham University, cosmic rays are still waiting to flood out of a gigantic bubble of gaseous material left over from a nearby exploding star which blew up more than 100,000 years ago. The Durham group reached the conclusion from observations showing that the cosmic ray intensity in the giant gas bubble is twice that observed elsewhere in the cosmos. Even more energetic particles (gamma rays), say the Durham team, making up a tiny amount of those hitting Earth's atmosphere, probably come from a neutron star circulating around another strange star, known as Cygnus X-3, some 30,000 light years away.

The dangers of the magnetic 'flip'

We saw earlier how virtually every celestial body is surrounded by a magnetic field. This, in a sense, is an invisible protective shell. The compass needle gives the clue. As it fluctuates throughout the day, it hints that currents of electricity are deflecting it. Electrically charged particles are attracted to the vertical path of Earth's magnetic flows, knocking off the negatively charged particles from the outermost layer of the atmosphere in the process. This layer soon becomes 'ionized' (hence the name Ionosphere).

The magnetic lines of force arc outwards along longitudinal lines above the Earth, and if a vast number of iron filings were sprinkled from a spacecraft just above the stratosphere, these arcing lines could probably become visible to the human eye as the filings oriented themselves along the force-field. Many particles are attracted to the northern latitudes of these arcs, situated between 76N and 79N. The magnetic field then acts as a trap or filter for the harmful cosmic radiation and high speed protons, as well as Ultra-violet (UV) radiation, most of which emanates from the Sun itself.

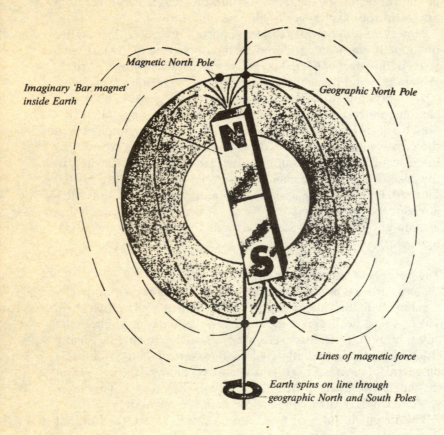

Earth's magnetic lines of force.

Source: *The Earth*, Granada Publishing, Steve Parker, 1985

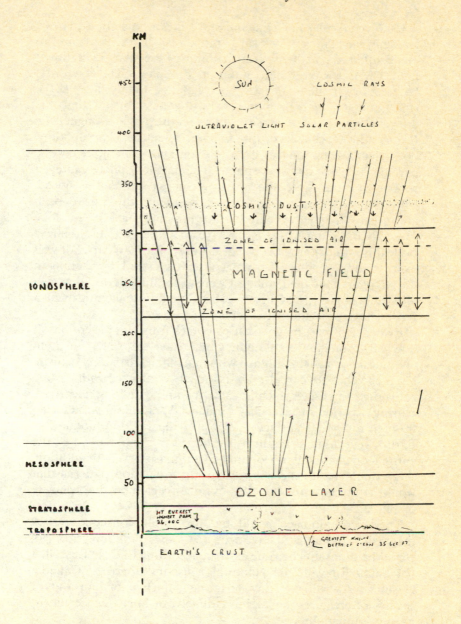

Earth's magnetic field and ozone layer, in different ways, repels much harmful cosmic radiation.

However, Earth's field, like the Sun's, is not constant, and has a tendency to fail at awkward moments. It varies from an idealization of a mathematician's notion of a regular gravitational field. The Earth, after all, is not a perfect sphere. The equatorial bulge, for example — the way the planet is slightly fatter at the sides, and its crags, mountains and low seas, means that this uneven distribution of density will affect the planet's magnetism. If it should weaken, according to some scientists, it would allow both excessive amounts of UV radiation and vastly more cosmic particles to reach our atmosphere. The magnetosphere can even reverse direction, with the north becoming the south. Indeed, within the span of geologic time magnetic 'flips' have been alarmingly frequent. There could possibly have been nine reversals in the past four million years. Other scientists believe twice this number have occurred, say every 220,000 years. For well over a century geophysicists have observed a steady weakening in the strength of the field, and if it were to continue it would vanish altogether in just 1,500 years time. But does this weakening mean a restlessness, or is another reversal on its way?

Evidence of past magnetic flips comes from samples of basalt dragged from the ocean floor, which have imprinted in them alternating weak fields aligned with that of the Earth's magnetism. Magnetic changes can also be intimately connected with cosmic collisions, intensive volcanic activity and gruelling Ice Ages. During a period of reversal so much UV radiation would seep through to Earth's surface that a great deal of the biomass — Earth's vegetation — could be destroyed in the process, and the entire food chain put in jeopardy as animals starved. In addition, there would be drastic changes in air circulation patterns that would continue until the field built up again in the opposite direction.

The overall impact would be dire in its consequences. A magnetic flip could well have occurred some 65 million years ago, when the dinosaurs became extinct, according to Dale Russell, a paleontologist with the National Museum Board of Canada. Cosmic rays would then have terminated the lives of thousands of species of creatures, either immediately, or later on because of harmful mutations.

During a reversal mutation rates are said to increase, thus causing a higher rate of fatalities among mutants, which die off before they can reproduce. The point is that, before a complete reversal, when the magnetic field dies away to nothing, a period of between 1,000 and 10,000 years could elapse. This is long enough to cause a severe distortion to parts of the world's biosphere and climate. Indeed, at the time of the reversal the increase in charged cosmic rays striking our planet could well be lethal to many species of plant and animal life. John Taylor believes that sharp changes in terrestrial magnetism account for many human ailments, including an increase in thrombosis.

Ocean temperatures are a key indicator. During the last 2.5 million years eight species of unicellular marine creatures still became extinct in droves after magnetic reversals, when cosmic ray bombardment was at its height. Many organisms contain magnetic grains used to orientate themselves, and a reversal could disrupt their orientation, leading to death. Often deep Antarctic seawater was unable to save them. Low magnetic fields, says Ian Crain of the Australian National University, could be the prime catastrophic causal agent in these extinctions.

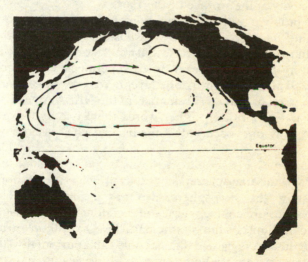

Changing magnetism is said to be related to Pacific Ocean temperatures in a two-way feedback; a rapid change in the magnetic field making the ocean dynamo act more strongly, and the ocean circulation itself sets up weak electric currents.

Source: John Gribbin, *Future Weather*, Pelican, 1983

There are, however, voices of dissent. Some scientists argue that it is not the magnetosphere that protects us from cosmic rays but the ordinary atmosphere. Its myriad of gas molecules in fact act as a buffer. Also, solar flares are thought to be unlikely to destroy ozone. Some astronomers, like Donald Goldsmith of the University of Berkeley, California, deny that the magnetic field has any great effect on cosmic 'rays'. If Earth's magnetism was to disappear overnight the radiation at the surface would rise by less than 10 per cent, which would be much too low to kill off entire species. In other words, it is not the field that is important, but the quantity of particles coming in. And a massive dose of radiation, says Goldsmith, is more likely to come from a supernova explosion with or without the field. Indeed, such explosions are more likely to be the real source of damaging cosmic rays reaching Earth. According to D.H. Clark and William McCrea of the Astronomy Centre at Sussex University, a supernova explosion would produce a hundred-fold increase in the cosmic ray count. In turn this would produce a thirty-fold increase in the mean level of radioactivity at the surface of the Earth.

One problem with this argument is the fact that the overwhelming majority of supernova explosions are too far away to do much harm to Earth. And those that are close enough happen only once every 100 million years or so. Supernovae close enough to destroy 50 per cent of the ozone layer occur only once every 2,000 million years or more, say scientists of the NASA Ames Research Centre in California. In other words there is only a 3 per cent chance that one occurred sometime in the past 66 million years.

The Magnetic Atmosphere
In a sense the atmosphere also has a magnetic field. The atmosphere is, of course, a gas, and therefore must consist, like all gases, of millions of sub-atomic particles. Over 100 years ago one Balfour Stewart suggested that not only can a current flow through the gases of the atmosphere, but, with its hurricanes and storms, it can also generate kinetic energy. Direct measurements of electric currents in the atmosphere are not easily undertaken; indeed they are well nigh impossible. Instead, ion current values at varying altitudes are frequently worked out from conductivity and electric field data by the use of Ohm's law. In any event it is known that the

atmosphere can actually drag the ionosphere across Earth's magnetic field to produce a primitive but powerful voltage. This means, in turn, that it must at some time become magnetically charged. Perhaps Earth's own field produces a magnetic effect which drives the electric part of the weather machine. Or perhaps evaporation of water from the surface negatively charges the Earth, but charges water vapour positively. In fact Volta, after whom the volt was named, in 1800 advanced an electrification theory based on vapourization and condensation of water to explain thunder.

One modern theory suggests that all thunderstorms recharge the Earth–ionosphere system. A competing theory is electro-chemical. This decrees that negative ions are captured by the land surface through some electrostasis phenomenon. What this means, in turn, is that when a planet the size of Earth is ventilated by ionized air it will charge up to approximately one million coulombs. But this is varied by the two-thirds surface water, and by winds carrying natural ions which are supplied at a steady rate by cosmic and radioactive decay.

Now, what happens when there are more sunspots than usual? John Nelson of RCA, who perceived the Sun to be a magnet, saw, in the course of his radio research, the relationship between sunspots and EM disturbances back on terra firma. When the Sun powerfully beams its lines of force out from the corona, about 100 trillion tiny particles inadvertently reach our atmosphere. What Nelson found was that when these particles suddenly increased they actually interfered with radio transmissions.

Of more relevance to Earth's weather patterns were the discoveries of Ralph Markson of the Massachussetts Institute of Technology. Making direct observations of the electrical potential of the ionosphere — way above the atmosphere — with balloons and rockets, he found that rays could well have a vital impact on the electrical condition of the atmosphere lower down. This might be because Earth has two hemispheric weather systems circulating in the north and south. If they could be seen from space with the human eye they would appear to have a dumb-bell type of configuration. Furthermore, they overlap remarkably well with the dumb-bell shape as we imagine it to be in the magnetosphere. Remember this can only be inferred, rather than proven. Once again we return to coincidental relationships of a curious kind. A

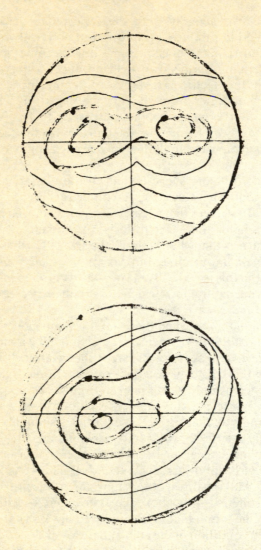

The top illustration shows Earth's magnetic field for the northern hemisphere for 1965.
The atmospheric pressure system for the same period shows remarkable similarities.

Source: John Gribbin, *Forecasts, Famines & Freezes*, Wildwood House, 1976, after J.W. King, *Nature*, October 1973

ridge of high atmospheric pressure is to be found near the northern force fields mentioned earlier. This ridge dominates the northern weather systems, so that any slight slackening of magnetic intensity can lower the pressure and cause marked changes in the circulation of the entire hemisphere.

The evidence seems to be growing that a more powerfully radiating Sun, and occasionally a weak geomagnetic field, can create more atmospheric depressions than usual. Walter Orr Roberts, a distinguished climatologist with the University Corporation for Atmospheric Research, is doing extensive work on solar ray bursts which could cause larger than average low pressure systems to form in the northern Pacific, and which could in turn decrease the intensity of the cold northerly air-streams. There is other evidence for large-scale outbursts being somehow connected to storm systems in the eastern Pacific heading towards Alaska. Depressions, of course, bring on rain, and a continuous stream of depressions, with their extensive cloud cover, could reflect back more solar heat and thus trigger an Ice Age if the phenomenon persisted for long enough.

One other intriguing explanation has recently been proffered. As we saw in the last chapter, the Sun has recently entered an interstellar cloud of hydrogen and helium particles. Paresce and Bowyer, writing recently in *Scientific American*, point out that the solar wind will only shield Earth from this cloud so long as its density does not rise beyond 150 atoms per cubic centimetre. Above this, and the hydrogen cloud will soon reach Earth's environment. Then it could react with the hydroxyl 'radicals' (one atom of hydrogen mixed with oxygen — HO). Soon, with an extra hydrogen particle it will turn into water vapour, create more cloud cover and depress terrestrial temperatures in the way I have described.

Sunspots and Ice Ages

The validity of the sunspot theory rests on the evidence yielded up by the Earth itself. They do seem to uncannily match the historical record of the Sun's fluctuations. Chinese scientists, indeed, are fortunate enough to be able to work from surprisingly lengthy historical climate records dating back some 5,000 years. A Professor Zhu Zezheny published a detailed study of different types of historical evidence (like changes in farming practices and prayers for rain), as well as more scientific data. The Chinese, after analyzing the data, found that a number of cold and warm spells coincided with planetary synods. The evidence from China has in fact greatly advanced sunspot/climate investigations in recent years. Cosmic particles, we now know, by breaking up nitrogen atoms, cause carbon dioxide to contain partly radioactive carbon 314 molecules. Trees absorb CO_2, as revealed in their tree rings.

Further studies done in the Bavarian Alps by Rudolph Reiter, found a definite relationship between Phosphorus 32 and Beryllium 7, two important elements that move down from the stratosphere to the troposphere. Reiter discovered that this radiation in the atmosphere was to be found two to three days after a solar flare. Rock formations also record climatic cycles. Australian scientist George Williams announced banded sediments containing familiar eleven- and twenty-two-year rhythms from 700 million year old rocks in south Australia. Bob Bracewell of Stanford University, California has used varve records going back over 1,300 years, and has found not only the eleven-year cycle but a longer one of 350 years, and yet another of 314 years. Researchers at NASA's Ames Research Centre in California, as reported in *Climatic Change* in October 1987, also report a varve periodicity of 10.8 years, interspersed with another longer cycle showing a lunar nodal tide.

A more complete record of the Sun's behaviour is logged in the Earth's giant polar ice caps. Edward Zeller of the Geophysics Department of Kansas University, for example, has traced billions of years of compacted snowfalls that contained nitrogen compounds. As man was not around to pump contaminants into the atmosphere, it was concluded that cosmic events were responsible for disassociating natural nitrogen particles, which were then washed down with the snowfalls. Tests in the early 1980s have now confirmed the efficacy of nitrogen-dating techniques, so that the way is now clear to determine whether it is cosmic rays or solar flares that cause the nitrogen compounds to form.

In the meantime there has been some more up-to-date evidence implicating sunspots. The drought in the Sahel, which seemed to end with the catastrophic floods in the summer of 1988 after nearly twenty years of aridity, began when the sunspot cycle was at its maxima. And 1988 has shown extremes of weather around the world. What about the phenomenal drought in the United States, where temperatures in 1988, and in 1986, were in the nineties for literally months on end, and which culminated in a much reduced grain harvest? The accumulating and disturbing evidence, from many astrophysicists and climatologists, is that solar factors are deeply implicated. Murray Mitchell of the US National Oceanic and Atmospheric Administration is just one who believes that droughts in the US midwest occur roughly a year or two after a double sunspot cycle.

Let us remind ourselves that Ronald Gilliland says that the Sun was at its 'biggest' in 1987. The National Oceanic and Atmospheric office says the Sun's activity could peak in late 1989. Bob Bracewell predicts another sunspot peak in 1991. Rhodes Fairbridge of Columbia postpones the period of maximum solar activity until the 1990s. John Shirley, based in California, says the Sun is now performing some unusual gymnastics with a kind of retrograde loop — actually travelling 'backwards' compared with the average direction it has been following over the past 1,300 years. He predicts an increase in volcanic activity and more climatic extremes on Earth. More scientists agree that the Sun has been more active in the past few decades than at any time since Galilleo. Indeed, as this book goes to press it is reported that scientists at the US National Oceanic and Atmospheric Administration have spotted two X-class solar flares — reputed to be the most violent of all. There is speculation that Earth could soon be ravaged by the most violent solar storms for 200 years, jamming radio communications and knocking out power lines.

Recent controversies concern what happened during the Little Ice Age, that prolonged cool period from 1645 to 1712 when there was a 'Maunder minimum', a period when remarkably few sunspots were observed. This has frequently led commentators to believe that sunspots are equated with a global warming, but the debate now focuses on the size and behaviour of the Sun itself as a measure of solar activity. Recently, French researchers have been arguing that the Little Ice Age was due to a drop in solar heat associated with a physical expansion of its outer layers. In February 1988 Leslie Morris, an astronomer at the Royal Greenwich Observatory, and other colleagues, denied this. They used old records of an eclipse in 1715, in which the size of the Sun could be measured by comparing the size of the Moon's shadow on Earth with known data. In short, then, logic prevails: a larger, more active Sun, as we have at the moment, means a warmer Earth.

The 1980s have produced a plethora of scientific papers on the new science of magneto-meteorology. Again, they were concerned with correlations, like the one reported in 1982 by Karin Labitzke of the Free University of Berlin. There seemed to be a link between the sunspot cycle and stratospheric 'weather' over the equator and in the polar regions. The correlation was particularly strong in regard to the west equatorial winds and the way they were likely to break down when sunspotting occurred, to allow the intrusion of warmer air. There were also strong correlations between atmospheric pressure over northern Canada. Indeed Labitzke and colleagues were able to correctly predict the weather in central eastern United States on the basis of findings.

Other scientists involved in magneto-meteorology can perceive relationships and suggest what the physical link might be. Hitherto, as we have seen, inferences have been made about electric charges and atmospheric pressure. Walter Orr Roberts was one of the first to perceive this, and his findings about the relationship with high latitude pressure changes and solar 'storms' goes back to the 1950s. Goesta Wollin, whom we met in the introduction, who predicted American snowstorms following sudden changes in the strength of the solar field, has also become a pace-setter. He is concerned with the linkage between the solar and Earth fields, but puts the emphasis on the latter. So far all his predictions have been about storms and freezing weather rather than heatwaves. Wollin believes the link is to do with the world's electrostatically conductive oceans. The atmosphere, after a day or two, becomes similarly charged.

Such explanations are plausible, but lack rigour. They are vaguely concerned with ionisation and with changing atmospheric pressure. They are not, we must remember, dealing with the more obvious effects of infrared radiation. What is becoming clearer is that long-term solar changes will affect climate, but not necessarily in the most obvious ways. The sunspot effect works through electromagentism rather than just through infrared heat. Russian research shows that the actual heat from the Sun peaks somewhat inconclusively when there is a moderate, but not a

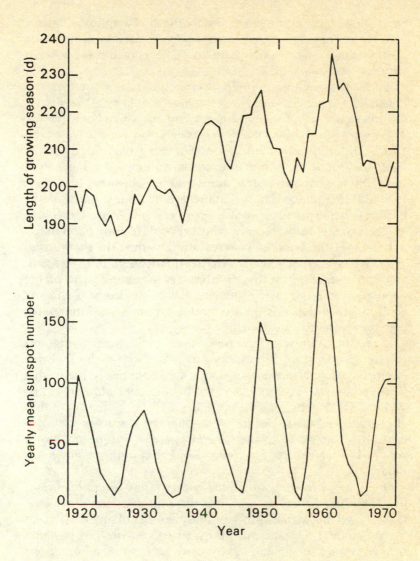

Evidence that the growing season at Eskdalemuir in Scotland coincides with sunspot peaks — hinting that sunspot maxima imply a warming.

Source: J.W. King *Nature*, 1973

59

great, number of sunspots. Some scientists, to complicate matters, believe the opposite occurs; when, for example, either a few, or a great number of sunspots seem to be responsible for marked variations in solar heat.

For the present it would help if one could declare categorically the state of current geomagnetic play — whether Earth's field is increasing or decreasing, in what way, and in which hemispheres. Then we might be better able to understand cosmic knock-on effects. Unfortunately the field tends to vary from one part of the world to the other, and from one epoch to another. Evidence from some 200 observation points across the globe shows that since 1930 Earth's magnetism has increased in many parts of the northern hemisphere. It looks, too, very much as if the geo-magnetic field has drifted westwards over the century, and that it has dragged the weather systems along with it. In those areas where it has weakened, such as in the Americas, the climate has in fact got markedly warmer. In the very northern parts of the hemisphere the increasing intensity has coincided with climatic cooling. Much depends on a complex syndrome of interrelated magneto–meteorological factors.

Determined efforts are now being made to fill the irritating gaps in our knowledge of the mysterious goings-on in the invisible electrically-charged nether world high above our heads. Pioneering work by climatologists at the Aspen Institute in Colorado, led by Walter Orr Roberts, prove that Earth's storm system is indeed changing. They also found solar-charged particles are con-centrated more in the northern hemisphere — Europe and North America — just where they are concentrated in the geomagnetic field.

One major problem with solar energy is the long time-lapses involved. Apart from the millions of years that convective heat takes to get up to the Sun's photosphere, sunlight itself takes twenty minutes to reach our planet's surface. And fast particles thrown out by solar flares take nearly twenty-four hours to get here, with more days passing before their influence reaches the surface layers of the atmosphere. This might explain why weather patterns, such as is manifested in the growing season, tend to appear one year *after* a peak in the cycle. Strictly speaking, however, a twenty-four-hour delay ought to seriously diminish any weather-shaping influences.

Now space scientists in the US, Europe and Japan are planning to embark on an ambitious joint investigation of Earth's EM field in the 1990s. The International Solar Terrestrial Physics Programme will send spacecraft to see how the solar wind originates, and to find out why the Sun oscillates on an eleven-year cycle.

Other satellites will measure the flow of plasmas striking the magnetosphere. A space probe will receive radiation signals such as X-Rays and UV rays unobscured by Earth's atmosphere or EM interference. Physicists will then take a giant leap towards a new understanding of cosmic plasmas and their effect on Earth's geosphere.

Hitherto, the compartmentalization of the physical sciences makes the practitioners so specialized that they become extremely wary of trying to integrate suggestive lines of reasoning that have a bearing on other disciplines. Geographers, to their credit, now admit the need for more integration. Conceiving the Earth as an energy machine ought not to be merely the prerogative of biologists. Analyzing the way energy is transformed and utilized could nowadays actually be synthesized into a study of the physical environment. In doing so it will inevitably be necessary to include the feedback effects of solar power, the rotation of the Earth and the not insignificant impact of human activity.

PART II:
THE VOLATILE EARTH

Chapter 4:
TRIGGERING
CLIMATIC EXTREMES

The climatic result of a cooling Sun, as postulated by the popular press in the manner described in Chapter 2, should be obvious enough. A loss of power of just some five per cent, indeed of just 2 per cent, would have appalling consequences for Earth. However, the idea that Ice Ages may be caused by a *hotter* Sun, a theory that was popular a few years ago, needs some explaining.

The Hot Sun/Cool Earth syndrome was the original brainchild of the astronomer Sir George Simpson. Ice Ages occur when the polar ice sheets are at their maximum in the northern hemisphere, and where the drop in global temperatures is caused by the build-up of glaciers on high ground, and snow sheets on the plains. All you need to get an Ice Age started, said Simpson, is more snow precipitation in the higher altitudes, and when this settles for long enough it solidifies into glaciers. It is the extra sunshine that will cause the additional evaporation from the oceans needed to cause the extra precipitation.

Could this strange theory possibly be true? Even a cursory knowledge of geology decrees that it could. In fact Simpson's reasoning, so simply stated in the foregoing paragraph, displays the kind of logic that most Earth scientists today use. He believed that, because of the angle at which the Sun struck the globe, the tropics would on average receive more radiation than elsewhere. There would then be a greater contrast between the tropics and the polar regions, and the weather machine during maximum solar output would be put into top gear. The flow of air would be speeded up, and there would be more evaporation than usual from the seas.

This geophysical paradox has been known for some time by scientists studying the Greenhouse Effect — the current global warming brought about by the increase in carbon dioxide being pumped into the atmosphere. Increased precipitation over cooler, northerly parts of the globe can actually cause the polar ice sheets to build up, rather than melt — at least if the warming is small.

The Hot Sun/Cool Earth theory is, however, flawed. It can possibly explain a northern hemispheric Ice Age, since extra snowfall alone will start it off. But some other mechanism would need to account for freezing temperatures in a southern Ice Age, since, as there is less land, the oceans would have to freeze solid to bring about the necessary drop in temperatures. Once the sea is frozen hard it gleams as white as the land-based ice sheets in the north, and perpetuates the cold by reflecting back the Sun's light and heat. What causes one kind of Ice Age is the cool summers and springs, rather than the extra cold winters, since the snow will last longer and ultimately turn to ice sheets. And as cool northern summers always coincide with severe winters in the south, it follows that a true global Ice Age is always accompanied by frozen southern seas.

Our complex, uneven Earth

Yet the virtue of the Hot Sun/Cool Earth explanation is the way it brings terrestrial convection currents into the picture. The Earth is somewhat like a giant domestic refrigerator: you first need heat to create the cold. It would be a bit easier for high-school geography teachers if the Earth were a perfect, unblemished sphere. Earth science would be a far simpler exercise if the planet were bone dry, like Mars, or completely covered with water to a depth of just a few hundred feet. It would be even better if Earth did not have an atmosphere.

Instead we have to reason with a complex planet whose surface is unsymmetrical and very uneven, being partly liquid and partly solid. Furthermore, it has a gassy atmosphere, a molten core and a shifting crust, all driven by internal heating or kinetic energy. And we must add the relatively new dimension of plate tectonics, which explains how the continents have drifted into their present positions. There are other complexities concerning how much of the Sun's radiation is reflected back into space, the circulation of the winds, and the dynamics of the hydrosphere — the way atmospheric moisture is finally returned, via rainstorms, to the oceans.

The volume of the ocean basins has fluctuated over time, and has had a great influence over virtually every other life-generating and sustaining force. It is now clear to Earth scientists that the extent of ocean water, its depth and shallowness, and the way it periodically invades the land and then retreats, has been a major climatic determinant. In fact scientists from the University of Rhode Island now believe that sea levels rose so high between 136 and 65 million years ago that the Earth's temperature fell by as much as 13C. It was clear that, based on carbon isotope analysis, so much low-lying land was under water at that time, with organic carbon being buried so rapidly, that atmospheric carbon was literally being taken out of circulation: hence the drop in temperature. So it is not only Earth's distance from the Sun that is important for living things, but the way the solid parts of the Earth are spread across the globe. There is a complicated feedback process that not only tends perpetually to moderate or reverse what we think ought to be the outcome of these physical facts of life. There is an important historical dimension, too, which implies that the feedback process — acting like an in-built governor — had different effects in the past. For example, the drifting continents, over millenia, must have brought about geomorphological changes that can themselves shape a planet's climate. Plate tectonics tells us that the climate, in other words, must have been different in the past because the topology of the Earth has since changed its form. Furthermore the rate of geophysical change may have been faster than applies today, distorting the rate of feedback in the process.

The idea that rigid land masses may be embodied in some kind of terrestrial conveyor belt (known as the plastic asthenosphere) came about gradually after the war, when submarine exploration using sophisticated remote-measuring devices proved the sea floor was much younger than the continents, probably no older than 200 million years, compared with continental rocks' almost four billion years. One enigma was resolved: the expanding crust does not mean that the Earth's circumference is expanding, so the rise and fall of sea levels does not imply that the oceans are periodically spread more thinly.

Nevertheless, the seas have often changed places with the land and sometimes they are so high that the world in a sense becomes flooded. Coastlines retreat seawards on wide continental shelves, cooling those regions previously subject to mild oceanic in-fluences, like Britain. As the seafloor spreads, so volcanically active ridges emerge that form shallow seas in the centres of oceans. This in turn displaces oceanic waters, and causes marine transgressions. During the Jurassic period, about 160 million years ago, parts of the world at present high and dry were all but submerged, including North America, Europe and Africa. On the other hand, when a new supercontinent is created there are fewer mid-oceanic ridges, so sea-level begins to fall. This could lead to the cooling of enlarging land masses, especially in the mid and high latitudes. The Permo-Carboniferous age is said to be the end-result of Pangea coming into existence.

But when land masses break up, or become mountainous, the cooling effect can become even more marked. The late Precambrian Ice Ages coincided with the breakup of an early supercontinent, with the Ordovician and Cenozoic Ice Ages occurring at a time of fragmentation. At other times, such as in the late Paleozoic Ice Age, the land has projected its rugged profile way above sea level. Then much more of the ocean was turned into ice along the world's littoral shelves. This type of ice age seemed to be a sequel to the Caledonian and Variscan orogenies (mountain-building epochs) in Europe, the Apalachians in North America, and other mountains. Similarly the formation of the Alpine chain seemed to occur prior to the last (pleistocene) Ice Age.

Hence all ice ages are, in a sense, cosmic winters, primarily spurred into being by changes in solar radiation. For when the earth was in an embryonic state, in the dim mists of time, the first ice must have occurred when a northern continental land mass drifted over the present North Pole. It was only during the Tertiary period, some 70 million to one million years ago, that the North Pole first became landlocked, and the South Pole ended up straddled across a mountainous land mass.

And then, each time the ice packs dissolved into the seas, a new catalyst for climatic change would take over. Water is a much better retainer of heat than dry land. The land would get hot during the day but cold during the night, as the infrared heat escaped back into space. The oceans, instead, would imperceptibly absorb the heat, and then, over the passage of hundreds of years, use it as a modifying influence on weather patterns, bathing the globe in warm, moist air even at the predicted beginning of a new Ice Age. For this reason much of the northern hemisphere is warmer than its latitude would dictate, because there is more land

to absorb the Sun's heat, and the moisture-laden air of the warm
Gulf stream of the Atlantic prevents temperatures from falling
excessively (71 per cent of the globe is covered by water; in the
southern hemisphere 81 per cent is water). In fact, global
temperatures would be considerably higher than they are now if
there was less land on Earth, as there would be more water to
retain solar heat. Temperatures would be at their highest — up to
12°C warmer — if all the present land-masses were concentrated
around the equator, with the oceans to the north and south.
Conversely, the Earth would be much colder if the oceans of the
Earth were somehow landlocked around the centre of the globe.
Even with solar energy constant, both hemispheres would be
locked permanently into an Ice Age of unparalleled severity.

From this perspective, then, it is the changing distribution of the
oceans, and not simply their 'heat engine' properties, that
determines climatic change. We know this is likely to be true
because of our combined understanding of continental drift and
the Earth's record of past climates. For example, during the
Oligocene period (32 million to 26 million years ago) Australis
rifted from Antarctica, while the land link between South America
and the Antarctic peninsular was broken. This led to changes in
the oceanic circulation of the southern hemisphere, because from
then on all the world's sea currents were interconnecting, and the
cold water around the Antarctic and the Southern Ocean
succeeded in cooling the surrounding areas. The closing of the old
gulf between the Arctic and the Pacific Oceans also cut off the flow
of warmer water from the equator, while it trapped the colder
water. The Labrador Sea was shifted further eastwards, and
Greenland ceased to be warmed by the North Atlantic currents.
Until 3½ million years ago there was no land link between north
and south America, and the raising of the isthmus of Panama no
doubt drastically changed the nature of the oceanic circulation in
equatorial regions. During the Miocene and the Pliocene periods,
the link between the Mediterranean and the Indian Ocean was
cut. The islands of the East Indies also rose from the sea, and this
reduced the exchange of water between the Pacific and Indian
Oceans.

For this reason meteorology is a difficult science, highly
resistant to any mathematical treatment, thus making its practitioners subject to the irritable criticism of the lay, sunshine-loving

public. The weather is only one manifestation of constant geospheric processes interlinked in such a curious, indeterminate way that predicting the weather to the accuracy of even 80 per cent frequently defeats the best computers. For example, Harry Van Loop of the National Centre for Atmospheric Research in Colorado says the 1988 US drought was 'teleconnected' with the low pressure systems that plagued northern Europe. Weather systems were 'stuck' in a stationary mode, and it is often the case that opposite extremes of pressure across the Atlantic occur, if only for the simple reason that high pressure systems cannot cover large portions of the Earth's surface simultaneously, a point emphasized by a spokesman for the Washington National Weather Service.

There is an often crucial and confusing time lag. Different air masses come either from the north or the south (i.e. they are either polar or tropical), or from the seas or from land masses (maritime or continental). Most of the precipitation east of the Rockies emanates from the Gulf of Mexico's maritime tropical airstream. The desert southwest of the US also gives maritime tropical air, although the mountains limit its influence to the west coast. The Pacific further north also gives maritime polar air, while Canada is responsible for continental polar air, especially in the north. But these air masses often take several days to work their effects through. And there is something even more fundamental adding to the complication that is described by physics. Momentum in an angular sense (for movement in a circular path) is related to the radius of curvature of any object in space. The Law of Conservation of Angular Momentum says that if you multiply velocity by mass and by this radius, you must always get a constant value. So any change in radius must result in a corresponding change in velocity.

High above the Earth's surface the fast-moving streams of air constantly interract with smaller-scale weather phenomena in such a way that some of the momentum is transferred down to weather circulations having a smaller radius. Hence, the velocities of these smaller circulations must increase if angular momentum is to be preserved. This is how tornadoes and fierce storms arise.

The Sun is the prime mover. Differences in temperature over equatorial and polar regions create energy through the unequal

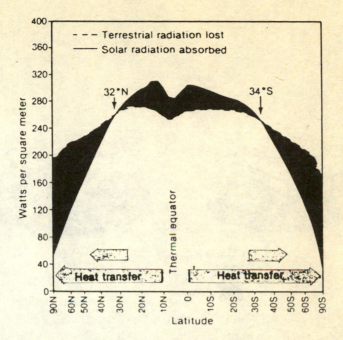

Comparison of the amount of radiation received and lost at various latitudes reveals that net cooling occurs poleward from 32° latitude. Heat is transported from the tropics to maintain an equilibrium.

Source: Joe Eagleman, *Severe and Unusual Weather*, Van Nostrand Reinhold, 1983

absorption of heat which is continuously transported polewards. Otherwise they would continually get hotter. Cooler air from the poles also moves southward. Water vapour evaporation means energy is absorbed, condensed, and becomes rain. Winds blow because of differences in both temperature and atmospheric pressure. In the tropics, with the large heat input, the air expands and becomes less dense: in polar regions the air becomes very compressed. This causes a stream of air to flow from the west to east over mid-latitudes at heights of about 40,000 fet — the jetstream. This explains the higher incidence of turbulent weather in the mid-latitudes.

The jetstreams, in effect, represent the larger scale priming the

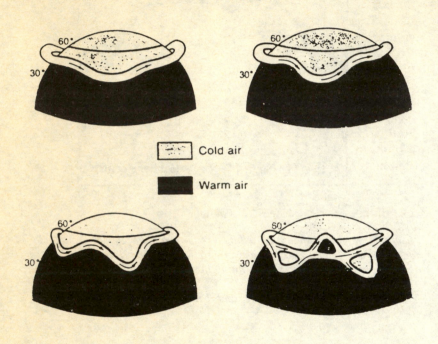

Cold air

Warm air

The jetstream may complete a cycle consisting of larger meanders until pools of warm and cold air are eventually cut off as the path of the jetstream straightens.

Source: Joe Eagleman, *Severe and Unusual Weather*, Van Nostrand Reinhold, 1983

smaller scale. The radius of a low pressure system (say, a frontal cyclone) can be about 500 miles, with individual thunderstorms of only about five miles, with a tornado of about half a mile. As the radii decrease, in other words, the velocities increase.

And it is the behaviour of the jetstreams, primed by the sun, that often gives the world its extreme weather. The streams should normally pursue a persistent westerly flow because of the Earth's rotation, but for some unknown reason buried within the complexities of atmospheric physics described above, the jet-stream sometimes slips and meanders. When it does this, it can bring on prolonged weather patterns. In 1976, for example, in both Britain and America, it developed a 'blocking high' — an anticyclone following a clockwise movement. At other times it meanders in a counter-clockwise phase, and creates persistent wet weather, especially in the mid-latitudes. This is why extreme, prolonged weather, in the absence of plausible meteorological explanations, is being blamed increasingly on the behaviour of the Sun itself.

The tilting, wobbling Earth
One great geological theory says that the Ice Ages emulate the rhythms and appearances of the changing seasons. A brilliant Serbian mathematician, Milutin Milankovitch, whose main works appeared in the years between the two World Wars, extended the seasonal changes in temperatures, caused by Earth's orbit round the Sun, across a much wider time span. The Earth's normal mode is circular and gives a more even spread of sunshine. But it occasionally goes from circular to elliptical, and then back to circular again. It is, however, the normal mode which implies lengthy cool spells.

It is therefore, according to the Milankovitch model, the growing eccentricity of the orbit that makes the climate warmer, thus perioically rescuing Earth from its Ice Ages. But when the orbit is more elliptical it means that, because one part of the orb is closer to the Sun at one stage, some months will become warmer than others. According to Milankovitch we are living in a fairly static time period, probably less variable than it has been for 1½ million years.

Climatic change, in short, can be determined according to Earth's attitude in space and the shape of its orbit. Gravitation plays a major role throughout. In ordinary times the tidal effect of both the Sun and the Moon pulls at Earth's equatorial bulge, which itself arises because of the centrifugal force of the planet's twenty-four-hour rotational period (should the Earth spin faster than it does the bulge would increase, and the Earth become yet 'fatter' at the sides as solid matter struggles to break free).

The Earth is at its closest point to the Sun in January, which makes the northern winters milder than they would warrant. Then the Earth accelerates in its orbit under the pull of the Sun, and when moving away from the Sun the gravitational drag slows earth down. The orbit thrusts the planet some 3 million miles further away from the Sun in July, in the process weakening the summer sunshine in the northern hemisphere to the benefit of the south.

There is another complication. Milankovitch reminded us that the part of the orb tilted towards the Sun naturally receives more solar radiation during the summer season. Over long periods of time, Earth's tilt varies as it rolls like a giant anchored ship in space, so that the amount of solar heat received alters in turn in the northern and southern zones. A smaller tilt of the axis means that both the northern and southern halves of the Earth get less Sun in summer and more in winter, producing less extremes of temperature globally. It follows that the more pronounced axial tilt will mean more extremes of temperature in both hemispheres.

This tilt cycle, according to Milankovitch, lasts for about 41,000 years. So every 20,000 years or so the Earth alters between being

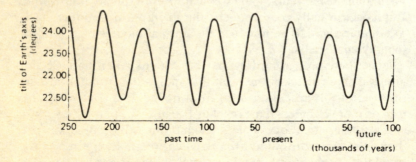

Variations in the tilt of Earth's axis over time.

Source: Fred Hoyle, *Ice*, Hutchinson, 1981

21.8 degrees from the perpendicular (or more upright) to 23.4 degrees (more tilted), and then, over the following identical period, tilts back again. The orbital modes do not change the total amount of heat arriving at the Earth's surface, since the average distance is always the same. Yet the most immediate effect is for heat to reach different parts of the Earth, to influence the temperature of the oceans and hence the global weather machine. Ultimately it brings on a cosmic winter.

In addition, because of the unbalancing effect of the gravitational pull on the equatorial bulge, the Earth wobbles like a top. This wobble lasts some 25,780 years, when the polar axis describes an arc in space as it continues its orbital journey (known as the precession of the equinoxes). This changes the direction of the tilt and emphasises the temperature contrast between the two hemispheres. In effect these movements bring about 'seasonal' changes of roughly 10,000 years each in length, corresponding to Milankovitch's renowned and hypothetical Great Spring, Great Summer, Great Autumn and Great Winter.

The Milankovitch cycle can also explain climatic rhythms for non-glacial times. The Earth's axis was more tilted 9,000 years ago, and the planet was closest to the Sun in July rather than, as at present, in January. Physicist John Kutzbach, writing in the July 1986 issue of *Nature* says that changes in the orbit changed the pattern of monsoons, and created a string of swamps and lakes in what are now North African deserts. Summers received 7 per cent more sunlight than now, and winters were correspondingly cooler. The important point is that orbital variations of the Milankovitch kind will bring about far more severe changes to the climate than solar variations themselves, although this has recently been disputed by researchers from the US Geological Survey. They say that several major climatic events in recent history, say up to 300,000 years ago, fail to tie in with the standard Milankovitch model, and suggest that variations in sunlight, ocean feedback factors etc ought to be given equal weight. This appears, at the moment, to be a renegade opinion. To most scientists it is clear that the 100,000 eliptical cycle is the most important, as the Earth is gradually pushed nearer the Sun, and then away from it; an effect that would largely override the other cyclical oscillations. And according to Milankovitch's own reckoning we are now pulling out of the present 'Great Summer'. And the next Ice Age, on this reckoning, will be with us in less than 40,000 years.

Proving the Milankovitch Cycles

During the 1930s and 1940s many geologists were won over to Milankovitch's radiation curves. The available geological strata had been used to test his calculations, and the theory seemed to fit. More recently Nigel Calder has published, in *Nature* magazine, orbital-dating calculations which seem to confirm the Milankovitch cycles, at least in regard to recently known Ice Ages.

Calder counted twenty-eight Ice Ages dating between 3.25 million and 649,000 years ago. This gives a mean period of 92,900 year intervals, close enough to the 100,000 year cycle. He says that the measuring techniques got better when scientists were able to study ocean-bed cores which revealed geomagnetic reversals. Hard evidence for these goes back to the last reversal about 730,000 years ago. Using his own techniques, and matching them with the 'core stage' magnetic findings of J.J. Morley and J.D. Hayes from 680,000 years ago, the other tilt and wobble factors seemed to be responsible for breaks in climatic cooling periods of roughly 30,000 and 60,000 years.

And yet there are some curious anti-Milankovitch anomalies that have still to be explained. One important anomaly was the discovery of a 25,000-year-old layer of peat in Illinois which could only have been formed in a warm climate. This finding contradicted the widespread belief that the period was then much cooler, culminating with the maximum extent of the Ice Age about 18,000 years ago. One German geologist said that some fossil molluscs he discovered on a gravel layer of an alpine river terrace dating back to the same era, are today only found in warm climes. Others also found winter temperatures of the minimum curves to be 0.7°C warmer than today when the winter prediction was supposed to be 6.7°C colder, and so on.

In 1976, however, important new findings rescued the Milankovitch theories. Scientists from the Lamont-Doherty Geological Observatory of Columbia University, after exhaustively studying oxygen isotopes of sea fossils, found curious cyclical trends. A litre of seawater would probably contain upwards of a million invisibly small creatures. Many of these marine organisms are miniature 'shellfish'. They have skeletons of two main types: Foraminifera (based on calcium carbonate) and radiolaria (with shells of silica).

Foraminifera accumulate in large quantities on the seabed. Scientists sometimes refer to them as 'chalk ooze', which ultimately solidifies into chalk. Radiolaria, however, forms a 'siliceous ooze', brownish in colour, which, over centuries, turns into a flinty kind of rock known as chert. These two types of sediment cover vast areas of the ocean floor.

The Lamont-Doherty team examined the fossilized remains from a length of core of sediment dragged from the seabed of the Indian Ocean. They checked the core backwards over 450,000 years and noted changes that were as distinctive as those in tree rings, and spelled out quite clearly to the trained scientific eye which were dry and which were wet summers. There were, for instance, traces of radiolaria that lived all through the half-million years being investigated, some of which flourish under warmer conditions than others. Changing ocean temperatures can also be monitored by checking the ratio of oxygen-16 to oxygen-18 atoms: water in the former molecule evaporates more easily than in the latter, hinting of wetter and snowier times.

The results of the core samples did seem to confirm the three Milankovitch orbital cycles of 23,000, 42,000 and 100,000 years. Furthermore, the Lamont-Doherty scientists hinted at a further progressive cooling over the next 20,000 years, possibly combined with a shorter-term trend over the next several thousand years.

Weighing the Ice Packs

One proof of the Milankovitch cycles arises from some simple physics. We know how much heat is lost from the atmosphere when it snows, and how much temperatures would need to rise to convert each gramme of ice back into water. Scientists can work out how much ice can be created and dispelled in accordance with changes in Earth's orbit; by the way increases and decreases in solar radiation can either increase or decrease the amount of terrestrial pack ice in the polar regions. Professor B.J. Mason, a one-time Director of the UK Meteorological Office, reckons that the amount of ice locked up in the last great Ice Age was 4.5×10^{22} grammes, or about eight million cubic miles. He reckons that there would have been an energy deficit from the Sun (i.e. the amount of solar insolation) of 675 times this, or 3×10^{25} calories.

This number of calories is remarkably close to that of the Milankovitch model, which suggests that the energy deficit for the last orbital tilt that brought about the last Ice Age was 550 calories. Hence, the changes in insolation in the critical seasons over the past 100,000 years correlated surprisingly closely with a tilt/wobble cycle that predicts the advance and retreat of the Ice Ages.

In addition to Mason's heat budget, there are the important radiation curves, arrived at independently of Milankovitch's calculations, that seem to point in the same direction. These curves, according to climatologists John and Katherine Imbrie, revealed earlier changes in summertime solar energy at latitudes of 55, 60 and 65 degrees North. When they reached their minimum ranges, they caused an Ice Age. The last three minima, they wrote, were bunched in groups of threes corresponding to the exceptionally cold epochs of 25,000, 72,000 and 115,000 years ago.

However, in view of the massive amount of terrestrial ice that is permanently with us (about 8 million cubic miles resting on various land surfaces), it is surprising how the balance is ever tipped towards a global thaw that would melt much of it. Certainly it is easier to melt ice than to create it, but when the orbit is nearly circular the temperate nature of the climate in both hemispheres is enough to make the glaciation deeper, since there are few hot summers to dispel the ice.

Milankovitch was confirming what we were discussing in the first chapter. The universe is dominated by rhythmic cycles of enormous variety. The seemingly changeless twenty-four-hour cycles of night and day is the most obvious, as is the diurnial tidal movement, and the regular rotation from summer to winter. The universe reveals even slower cycles: for instance, planets are said to die off and be created over a massive cycle of 80 billion years. We have seen how the sunspot cycle averages just over eleven years, although it can go up to seventeen years. This seeming irregularity seems to upset the cyclical rhythms of Earth's biosphere. Heavenly cycles, in short, can drastically alter the global heat balance and the 'weather machine', and make it far harder to determine how the climate will be affected. The Sun, the tides, terrestrial wobbles and tilts, the eliptical orbit, all play their part in the intricate web of cause and effect. All this is difficult enough to unravel, but it is made doubly difficult when the mechanism is sometimes reversed, so that one is unclear whether the effect precedes the cause.

This last observation brings us to the next chapter. We have so far been discussing atmospheric leverage and orbital movements, and their effect on climate. The spin of the Earth is, however, an even more important factor in inducing Ice Ages, according to some theoreticians, than is clear from what has emerged in the foregoing narrative. It involves, as we shall see, a complicated feedback mechanism in which the spin of the Earth, its atmospheric circulation and volcanic eruptions, all seem to be peculiarly interconnected.

Chapter 5:
THE VOLCANIC CONNECTION

At the beginning of Chapter 3 we were discussing the tidal influences on the world's oceans, and the likelihood of serious floods arising from an Earth–Sun–Moon alignment. At this point we should remind ourselves that if the Earth was entirely covered with an even envelope of water, changing tidal currents would hardly be noticeable.

The fact is, however, that a sizeable 30 per cent of the Earth is solid land, which succeeds in breaking the path of these tides. All this creates frictional resistance which, over centuries, has actually slowed down the Earth's rate of rotation. In other words, the day grows continually, if imperceptibly, longer, and has been doing so ever since the Earth was first formed. The evidence for this comes from the fine ring-like banding found on some reef-building corals, which can be counted rather like tree rings. From studies done by C.T. Scrutton of the British Natural History Museum and J.W. Wells of Cornell University, it has now been established that coral shells have markings that can measure years, months and even days.

Scientists have now determined directly, by bouncing laser beams off a mirror left behind by the Apollo astronauts, that the Moon is receding from us by about one inch a year. So they have been able to calculate that Earth's day is increasing at a rate of one second every 62,500 years, or sixteen seconds every million years. In the last 400 million years, assuming the rate of increase for the moment to be constant, the day has lengthened by 6,400 seconds, or almost 1.8 hours (we must also assume in our calculations that the Moon was captured very early in Earth's history — and we will have more to say about the Moon in Chapter 7). In other words, 400 million years ago the day would have been 22.2 hours long, and the year 395 days long. Indeed, working backwards with a pocket calculator one can assume that the embryonic hot Earth, formed some 4.6 billion years ago, was spinning so fast that the day was only 3.6 hours long.

The concept of the changing length of day (LOD) is of great importance in our discussion of climatic change, but not for the most obvious reasons. Writing in *Earth's Earliest Biosphere*, earth scientist James Walker points out that the faster rate of rotation in Precambrian times (i.e. when the day was shorter) had an important bearing on the early climate and biosphere. The primary effect was the modification of diurnial temperatures. Climate would have been more equable because a particular spot on Earth would have less time to warm up during a shorter day, and less time to cool off during a shorter night.

But there is another way in which the changing LOD can affect the Earth's climate. Tidal influences can actually pull on the solid surface of the land, in spite of earth substances being held together by strong molecular forces. The planet literally bulges at the sides, and crustal heavings take place deep in the interior. The law of conservation of energy tells us that these tugs and strains are converted into heat, so as the Earth loses fractionally in its rate of spin it gets marginally hotter. Not only that, but Earth's innards have changed during its evolution — by expanding or contracting and by redistributing its mass — and this too could have affected its LOD. In technical terms this means there are occasional changes in Earth's moment of inertia.

The conclusion that many earth scientists are now drawn to, in the light of this knowledge, is that both solar and lunar tidal forces can precipitate seismic events deep within the Earth's crust. For example, in a paper published in *Nature* in July 1983, S. Kilston and L. Knopoff of the University of California referred to evidence of a lunar-quake link. Concentrating on a region in southern California, their examination of seismic records revealed a not insignificant 12-hour lunar fortnight and an 18.6 year period relating, respectively, to the positions of the Moon and Sun, and to seismic shocks in the region. In other words the quakes and tremors in the area *always* seemed to occur on a fortnightly or 18 to 20 year cycle. The Sun and Moon, wrote the authors, are not necessarily responsible for all the mini tremors in the area. But they were strongly implicated, and this is the important point, in the bigger quakes.

Other rigorously formulated evidence implicating the Sun with earthquakes comes from Russia. Soviet physicist A.D. Sytinsky of

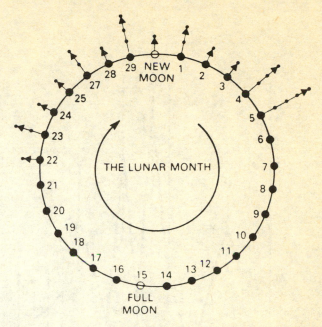

Twenty-three major earthquakes and volcanic eruptions distributed by lunar phase. All twenty-three are found at or near the New Moon; the statistical odds against this 'clustering' effect are 10,000,000:1.

At New Moon, the Sun and Moon are in conjunction, and their gravitational forces are combined.

Source: *Judgement of Jupiter*, Richard A. Tilms, Book Club Assocs, 1980

Leningrad's Institute for Arctic Research, has exhaustively analyzed data on 594 quakes of the magnitude of 6.5 or more. He found that energy levels peaked about a year after the solar maximum had been reached, and when the central meridian region of the Sun was particularly active.

Other evidence was more circumstantial, but intriguing. In December 1972 the Nicaraguan earthquake, which took the lives of 12,000, was suspected of having been triggered into life by a noticeable shuddering of the Earth as monitored by seismologists in the preceding months. In 1976, 1982 and 1985 (at the time of the Mexican quake) the changing length of the day seemed to play a

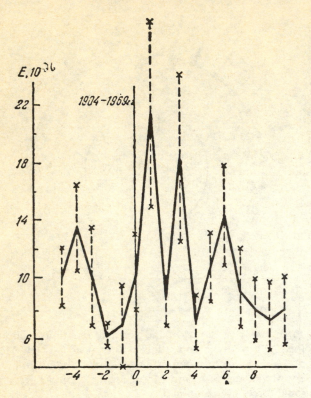

Professor Sytinsky's chart of Earth's overall seismicity shows a marked peak shortly after a solar maximum.
The vertical line is expressed in units of 10^{26} ergs, and the horizontal line denotes years before and after the solar maximum.

Source: *Cycles of Heaven*, Playfair & Hill (1978)

part, and other measurements suggest that the behaviour of the Sun was involved.

But how could earthquakes affect the LOD? Scientists believe quakes can make the Earth more compact, which in turn makes it spin faster. As we have seen, a spinning object becomes more compressed, and its rate of rotation must increase to keep total angular momentum unchanged. In 1988 Benjamin Fong Chao of the Goddard Space Institute, and colleagues, published data on the precise positions of satellites whose orbits were highly sensitive to Earth's rotation speed, and to the orientation of the axis and ellipticity. They then looked at the timing of about 2,000 earthquakes, which they said accounted for about one per cent of the total of the change in LOD.

The fact that Earth's rotation is not constant was first established more than half a century ago when the movement of the stars across the sky showed a baffling, if exceedingly small, irregularity. But it was a French astronomer, A. Danjon of the Paris Observatory, who first made the startling claim that has stimulated an entirely new scientific investigation. He spotted a change in the length of day in 1959, another year famous for an unusually big solar flare. This he did by subtracting Universal Time (UT) from atomic clock time (AT), which also coincided with a sudden drop in cosmic radiation.

The discoveries of scientists like Danjon later led to the

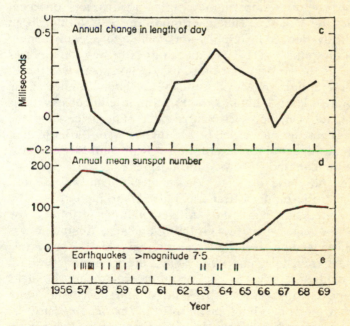

A comparison of the annual mean rate of change in the length of day and the annual mean sunspot number. It can be seen that earthquakes occur when there is a minuscule speeding up in the length of day.

Source: John Gribbin, *The Jupiter Effect*, Fontana, 1977

85

formulation of the Jupiter Effect, the subject discussed in Chapter 1. Those who had been dismissive of the Effect, like Don Anderson of the California Institute of Technology (Caltech), were obliged to have second thoughts. Ironically he himself had garnered more than enough evidence to prove that seismicity has an eerily close relationship, not with volcanism, but with tiny changes in the speed of the Earth's rate of rotation, or, in other words, in the changing length of day. As we shall see, seismicity and volcanism seemed to fit in with the relationship almost as a by-product.

Now, one of the embarrassing features of the Jupiter Effect was the non-existent San Francisco earthquake, prophesized for 1982. This seemed to cast aspersions on the entire theory. And yet the predictions of Drs Gribbin and Plagemann were only a couple of years out, since quakes did in fact occur near Los Angeles in 1979. Curiously one of the most famous volcanic 'disasters' of recent times was the Mount St Helens eruption in 1980. John Gribbin admits he overlooked its implications. With a force equivalent to 100 million tonnes of TNT (10 megatonnes) it was 500 times as powerful as the Hiroshima bomb. In addition there was the El Chichon eruption in Mexico two years later. Studies using ground-based lasers, aircraft sampling and satellite observations showed that El Chichon spewed up a cloud of particles across the Pacific as dense as that ejected by Mount Agung in 1963, which in turn was the biggest since the eruption of Katmai in 1912, and was predicted to cool the globe by half a degree.

Bearing all these intricate trigger mechanisms in mind it seemed that the Jupiter Effect was indeed working out its early 1980s prophecy, but in a slightly different manner. Gribbin chose 1982 as the year when the Grand Planetary Alignment was supposed to take place. In that year a remarkable 180 year cycle would have come to an end when, at last, the outer planets would have caught each other up. Solar output would then be at its maximum. Gribbin now realizes he should have paid more attention to the world's hot spots. Writing in *This Shaking Earth*, he says that nineteenth-century volcanic activity reached a peak on the graph in the decades prior to when changes in the length of day were observed. It was time to bring volcanoes into the picture.

The Hot Earth

The world's press began to itemize quakes with increasing frequency. There was a solar minimum in 1976 which coincided with the great Guatamala shock in which 23,000 people died. California's mammoth Lakes ski resort centre east of Yosemite National Park had for some reason been jolted by a series of quite strong tremors ever since 1978. Furthermore, in 1982, US seismologists suspected that volcanoes would soon erupt. Alan Ryall, of the University of Nevada, predicted a spate of volcanic-type seismicity. The reason, he said, was that the small quakes were taking place much closer to the surface, probably as a result of the slow upward movement of a hot river of magma. Other seismologists noticed more hot-spot activity. Fumeroles (steam vents) were beginning to emerge in earthquake-prone regions.

It was the fumeroles, of course, that first gave life to the barren world that preceded the fecund and variegated one we now inhabit. At the beginning, when the interior of the Earth was in turmoil, the concentration and aggregation of matter through gravitational attraction from the centre outwards turned the gelling planet into a series of hard, concentric shells. The innermost core became the densest, attracting the heavy elements such as nickel-iron.

Earth is indeed hot deep down, but the flow of terrestrial heat is small compared with that from the Sun. Yet the cumulative power of Earth's heat, resulting in fiery streams of magma bursting to the surface, is tremendous. Recent guesses of the temperature at the centre are about 12,000°F — more than 4,000 degrees hotter than anyone suspected. The molten core is a product of the radioactive decay of matter, and the crust itself consists of constantly degrading radioactive minerals. The outer layers and crust of the primordial Earth were left, as a result, with the lighter substances such as silicon and aluminium.

These days a knowledge of radioactivity provides us with a deeper insight into the depths of the Earth, hitherto the domain of poets. And a knowledge of what goes on down at 'middle Earth' can help us understand what goes on, not just at the surface, but in the atmosphere. When, about eighty years ago, Lord Kelvin made one of the first scientifically-based estimates of Earth's age, by calculating how long it would take a proto-planetary body to cool from an initially molten state, he was way out by several orders of magnitude too young. Now the generation of seismic waves by earthquakes gives a sharper picture of what the interior of the Earth was like. The image of the golfball emerged — a descending series of spherical cones. First came the crust, recognizably solid with pockets of oil, gas and magma. Then came the mantle, about thirty miles below, where 2,000 miles of heat and gravity could compress crystal. Then came the 1,000 mile deep outer core, an inferno of molten iron compounds. The 800 miles at the innermost centre was thought to be a solid red-hot ball unable to liquify because of immense pressures.

This picture, in 1988, is somewhat out of date. Using sophisticated new tools for drilling, and computer mapping of the deep underworld, it appears that temperatures at the core, as we have seen, have been massively under-estimated. There is, first, the new picture emerging from the presence of helium-3 which bubbles to the surface. The varying concentration of helium-3 at different sites hints that the mantle is layered in a number of sub-regions with asymmetrical boundaries, but which are all somehow interconnected. Other information comes from computerized axial tomography (CAT). The vibrations emitted by quakes tend to slow down or speed up, depending on temperature and the density of the interior regions they pass through. It seems that the internal boundaries are far less rigid than the upper ones. There seems to be a hot metallic 'sea' upon which there are analogous 'continents' floating, reflecting the drift that goes on at the Earth's surface. According to Jeremy Bloxham of Harvard, there are several pairs of 'hot spots' at these fluid boundaries, producing magnetic anomalies similar to those of sunspots. As these spots seem to have grown bigger by 10 per cent during the past 150 years, says Bloxham, we can expect a magnetic reversal in about 2,000 years time.

The volcanic connection
As we saw in Chapter 1, the tension between the gravitational pull of the large bodies and centrifugal force keeps the planets and moons in their regular orbits. The same principle applies to Earth's atmosphere. What prevents the oxygen and nitrogen particles from falling to the surface is the convection heat in Earth's warm, equable climate. Later on, however, it was the volcanism that produced a mixture of gases from inside the hot

Earth — carbon dioxide and traces of nitrogen and sulphur that formed the air we breathe. And it was the water vapour, created from the moist volcanic gases, which lowered the temperature and allowed the first rains to fall. It was also the rains, of course, that germinated earthly life after the hydrosphere was formed.

Soon after, distinct zones of climate appeared, which meant that from then on Earth's temperature would not always be constant, or even warm. During the cooler solar periods atmospheric gases had a tendency to fall back to the surface.

The planet's dominant characteristic — one that features prominently in both its physical and biological make-up — is the element carbon. Our world is, in effect, a carbon recycling entity, and its main climatic and biological features are determined by this, as we have seen when discussing the relationship between sea levels and temperature. The early CO_2 that was in the atmosphere drifted back to the surface — more particularly its oceans — and ultimately fused with the ocean floor sediment. When the sea floor sank beneath the edges of colliding tectonic plates, the carbon became buried even further into Earth's interior. Ultimately the carbon molecules would get forced back to the surface again when volcanoes erupted explosively into the atmosphere. According to James Walker and Paul Hays at the University of Michigan, it is this carbon recycling phenomenon that accounted for both the unusually warm epoch during the Cretaceous period, and for the warmth of the primordial Earth under a Sun that was much dimmer than at present.

In the meantime the atmospheric blanket was just right to keep the surface temperature between the melting point of ice and the boiling point of water. This median temperature allowed water to flow, the oceans to swell, and would also have dissolved some of the CO_2, thinning it slightly in the oceanic 'sink'.

So it all started with volcanism, and it was the unstable land-masses that gave rise to both volcanic eruptions and the presence of carbon dioxide in the atmosphere. The continents, runs post-Wegener wisdom, came together as one land-mass after having split up and reformed several times before. They drifted apart again as the seabed spread and pushed up from the edges of the continents.

According to researchers at Caltech, about 400 million years ago Gondwanaland straddled the South Pole, where it was being lifted up by a rising heat-spot in the mantle. The excess mass at the pole destabilized the Earth's rotation to slowly tilt the entire mass, so that 100 million years ago Gondwanaland was at the equator. Then, stretching to fit the Earth's equatorial bulge, Gondwanaland was torn apart by rifts, one of which became the mid-Atlantic ridge. The continents then assumed their present positions.

But the surface had to stretch and contort. Mountains occurred where one plate was forced beneath another, or where they pushed the seabed apart and shoved out new material. At other times the colliding plates threw up entire mountain chains, such as the Himalayas, or the deep trench that was created long ago along the Pacific coast of South America.

As the plates continued to move imperceptibly, rolling currents of hot mantle-rock are thrust aside. Millions of tons of rocks become dislocated, and thrust further along fracture planes, called faults. These fault zones are prone to burst into activity because of the way in which the slabs push relative to each other, and by the way stresses and strains are built up over a remarkably thin skin. Frictional heat is also generated by grinding plates, and it is this which produces quakes. In fact geologic instability implies frictional heat, as the tectonic plates bump and grind together, melting granite and making the crust thinner and hotter.

One important clue has come from earlier studies showing that partially melted rock from the hotspot not only releases itself up through the crust, but actually spreads laterally beneath it. The plates build up tremendous pressure, and melt rock to form magma. Some 80 per cent of volcanoes are created in this way. So it is easy to see how volcanoes can be provoked into becoming fiery monsters. According to Peter Vogt, a marine geophysicist at the Naval Research Laboratory in Washington, volcanoes are the product of a growth of convection activity generated below the Earth's crust. There is a continual adjustment to this bending, malleable flow. Finally the crust gives way violently, releasing unfathomable hydrogen-bomb-like shock waves in places like the Pacific Ocean ridge, still thrusting the sea floor eastwards. Similarly, seafloor spreading on the western edge of North America — the same geophysics that brought the Rockies into being — also expand ridges over hotspots where smouldering volcanoes sit, like Mount St Helens.

Strangely, it is water that gives volcanoes their explosive power,

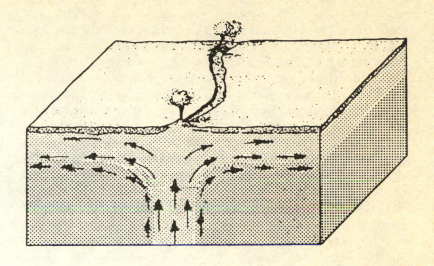

Erupting volanoes as a product of spreading ridges. Rising convection currents add material to the sliding plates that make up the sea floor, driving them apart.

Source: G.A. Eiby, *Earthquakes*, Heinemann, 1980

as the tremendous heat of the magma creates pressure which violently blows out the plugged andesitic vent at the top. The escaping steam and gas shatter rock and blast magma into the air. Some explosive eruptions are so great they can exceed the energy produced by a 10 megaton bomb, blasting out much ash, debris and hot lava: some volcanoes eject less lava and more of a mixture of gas, usually sulphur dioxide, hydrogen sulphide and carbon dioxide, plus dust mixtures. Lake water accumulating over a long-dead volcano can also trigger them violently back into life. Such hotspots may stay dormant for centuries, but they can form a massive network of chambers and fissures as molten rocks subside deep below. As the rocks cool, they emit huge amounts of gases, mainly CO_2, but also sulphur dioxide, and possibly cyanide. In the usually placid waters of Lake Nyos, in the Cameroon, an eruption of volcanic gases in August 1986 took the lives of 1,200 people.

The Earthspin factor

But it is the ash that is the major problem. Volcanoes can actually change the weather by spewing out great plumes of particulate matter, thus partially blocking out the Sun's heat and light. This can be particularly effective when fine grains of sulphuric materials and silicates smaller than a micrometre (10,000th of a centimetre) rise as high as the stratosphere. The thin air up there is very dry, and rain is unable to wash the aerosols out. They can take — as in the case of Krakatoa — up to a year to fall back to Earth.

Thus we can formulate a new mechanism to explain periodic episodes of Earth cooling; one that, as we saw at the end of the last chapter, entails a two-way cause and effect method, where the effect sometimes becomes the cause. Let us take the case of Krakatoa again. The Krakatoa event of 1882 could have been an Earthspin trigger, as the dust-laden atmosphere was so overladen it jolted the Earth, which then precipitated more seismic activity.

John Gribbin agrees with this reasoning, saying that volcanoes are part of a 'complex feedback mechanism'. But he still puts the emphasis on the Sun. It is the Sun, after all, which alters the atmospheric circulation, and which in turn changes the spin of the Earth, which triggers volcanism. The last link in the chain, then, is the volcanism itself, which exerts an additional effect on the atmosphere. Supporters of the Jupiter Effect are even claiming that it was probably the Sun's influence on the spinning Earth that provided the extra dynamic that precipitated Mount St Helens and El Chichon, and scientists at the University of Michigan have found that the volcano La Soufriere in the Caribbean erupted within a ten-day period coinciding with the fortnightly spring ocean tides (when the tides are at their maximum). The odds against this happening by chance were reckoned to be less than one in ten.

Other scientists found that the late 1950s, a period noted for seismic activity, coincided with a maximum lengthening of Earth's spin. A Dr R.A. Challinor found that length of day (LOD) changes, when subtracted from other regularities, followed precisely the change in the cosmic ray flux. When the Sun is more active, the Earth slowed down. He also noticed a curious increase in the LOD over the fifteen years from 1955 to 1970, although the rate of change varied, up and down, by a percentage point of a millisecond. Gribbin himself mentioned the massive 1972 solar 'storm' that coincided conspicuously with a slow-down of more than 10 milliseconds, the biggest known in one day.

A short while later, confirmation of the link between the sunspot cycle and the changing LOD came from Bob Currie's high-speed computer. Currie is a mathematician once employed by the NASA–Goddard agency. He was able for the first time to sort out space signals that fluctuated regularly, from the ones which stood out from the more random background mush such as radio noises. In 1980 Currie found unmistakable evidence of matching solar cycle signals.

This evidence also backed up Challinor's findings, which confirmed Russian evidence of maximum and minimum sunspot activity levels. Seismicity, noted Challinor, occurs at both the solar maximum and minimum cycles. True, the idea of an extra millisecond on the length of the day may not sound too exciting, but when measured in conjunction with the entire weight of the Earth, at 6×10^{24} kilograms, it has to slow down immediately to make the shortened day possible.

However, as the link between LOD and seismicity became better understood, new explanations began to be proffered. It is the acceleration and braking, in a sense, that triggers the quakes, rather than any gradual changes in LOD. So when the Sun or planets are rising or setting, and the spin of the Earth is subject to juddering caused by solar irregularities, the shear forces will be at their peak. Hence the mantle can be suddenly fired into movement from its hairtrigger balance.

Then, to increase the explanatory power of the LOD/seismic theory, the atmospheric leverage mechanism was brought into the picture. Disturbed weather, not necessarly caused by volcanic ash particles, could itself slow down the spin of the Earth. As the winds whirl across the land, encountering differently heated pockets of air, Earth's momentum can be sped up or slowed down literally by wind friction, reducing the length of day by some 20 milliseconds or so. One theory, from the Russian geophysicist A.D. Sytinsky of the Institute of Arctic Research in Leningrad, suggests that solar flares slow down Earth's spin by whipping up cyclones.

One attempt by scientists to see by how much the rotation was affected by weather patterns was conducted by meteorologists at the Australian National University at Canberra. Kurt Lambeck and Peter Hopgood calculated from wind speed data going back twenty-two years (roughly two sunspot cycles) the variations in the speed of circulation of the atmosphere. They worked from the premise of the angular momentum of the Earth. Hence, the faster the winds turn, the slower does the planet. After a comparison and a simple subtraction had been made, the measured fluctuations in rotation could be worked out. When the Sun and Moon's tidal effect had also been subtracted, Lambeck and Hopgood discovered an irregular but distinct rotational fluctuation.

The cyclical, reinforcing nature of Earth processes was clinched earlier in 1980 when new evidence emerged of the way in which the atmosphere can alter the length of day, if only by thousands of a second or so. A group of scientists from the National Centre for Atmospheric Research in Denver, and from the Meteorology Office in Bracknell, Berkshire, succeeded in adding together parcles of air pressure and wind speeds gleaned from the global Atmospheric Research Project, a satellite-based study.

More recently, in 1988, scientists have shown that short-period wobbles of the axis, in particular the polar motion, are mainly caused by changes in angular momentum of the atmosphere at the equator. As with the discovery of earthquakes making the Earth more compact, these new findings are made with the aid of astronomical radio sources and satellite laser ranging (SLR) in conjunction with orbiting spacecraft.

The logic of the new Ice Age law, then, dictates that it is the regions of geologic instability that are naturally most at risk from the changing spin of the Earth. The argument runs that any change in rotation could tickle highly sensitive regions into life. And more sudden changes in LOD can result in a more noticeable jolting and juddering.

But the stratigraphic evidence of the rocks hints that quakes and eruptions have been triggered many, many times in the past when geologic instability was much greater. In the old days, it seems, volcanoes blew their tops with much greater violence than they do today. Two important things we know already: the rate of physical evolution of the planet has not always been constant, and the shifting tectonics causes volcanoes to erupt. So when earth movements speed up, the slow creep of the tectonic plates that hitherto provided a regular movement of six centimetres can suddenly push things along at a regular six metres, and set the ground rumbling.

Further, it is at periods of increased geologic activity — as in prehistoric times — that seismicity and mountain-building seem to be on the upturn, and the sideways drifting of the continents seem to be on the increase. Everything seemed to be writ large in those times, and this applied also to the violent swings in climate. If volcanoes have a tendency to cool the climate, then millions of years ago they cooled with a vengeance.

Chapter 6:
HOT ASH — COOL SKIES

There is one contentious issue remaining, that will be the subject of this chapter: could volcanoes, however large and however violent, eject enough ash into the atmosphere to significantly cool the Earth? The answer, although it is a tentative one, must be Yes — because the scientists say so.

A study by Mick Kelly and Chris Sears at the Climate Research Unit at the University of East Anglia, shows that of four great eruptions over the last 100 years temperatures in the northern hemisphere were reduced substantially in the first sixteen months after the eruption. This was backed up by studies at the University of Urbana, Illinois. Ten major volcanic explosions since 1869 were recorded to have decreased dramatically Pacific Ocean temperatures shortly after each event.

In recent years Hubert Lamb, Britain's foremost climatologist, has been busy developing a Dust Veil Index (DVI) as a yardstick to measure the sun-obstructing qualities of dust hurled up into the stratosphere. Those rating the highest DVI were defined as being the most climatologically important. Now, most of history's sky-darkening eruptions have been chronicled, and the list is a long and disturbing one.

Kevin Pang from NASA's Jet Propulsion Laboratory (JPL) in Pasadena, and James Slavin from Goddard in Maryland, report that a famine that wiped out half of China's northern population in about 205BC may have been caused by the effects of an erupting volcano in Iceland. This was worked out by noting the levels of sulphuric acid deposited in the Greenland ice core. Sulphur dioxide would have mixed with water vapour in the upper atmosphere to form sulphuric acid droplets or dry fog.

There are many other recorded events in which climate seems to have been involved. Ancient Chinese chronicles from the Han dynasty tell of crop failures and massive starvation in an unnaturally cool and wet summer. In 536AD Rabaul, in what is now New Guinea, erupted to blot out the sun for up to eighteen months in Mesopotamia. Over 200 years ago Laki in Iceland disgorged the greatest volume of lava into the environment anywhere in history. Lava filled two river valleys, and covered more than 500 square kilometres. Stunted grass and sulphur poisoning from the accompanying gases (similar to the occurrence in the Cameroons) killed most of Iceland's livestock. Benjamin Franklin, living in Paris as the first US ambassador to France, wrote of the 'constant dry fog' all over Europe, and suggsted that the Laki eruption was to blame.

The Tambora blast in 1815, in the Dutch East Indies, was the second largest known volcanic explosion in history. In November 1815 the average temperature of central England had dropped 4.5°F, and continued to be well below average for two more years. Climatologists rank Tambora as the greatest producer of atmospheric dust in the last 400 years. And when neighbouring Krakatoa erupted in 1883, French astronomers in Montpelier observed a 20 per cent decrease in solar radiation for three consecutive years. It poured some 50 million metric tons of ash into the skies, creating great drifting clouds of vitreol.

There have been numerous, and massive, eruptions in the twentieth century: Katmai in 1912, Mount Agung in Bali in 1963 (reckoned to have reduced solar heat in the Soviet Union by up to 5 per cent), Taal in 1965, and Mayon and Fernandina in 1958; the dust from Mount Katmai in Alaska in 1912 was enough to lower the temperature of the whole globe for nearly three years, vast quantities of smelly ash and gases fell back to the surface to destroy crops in Caithness, Scotland, and the June Sun was said to be so weak it could barely be seen glimmering through the haze even in southern France. The active periods of all these eruptions seem to have occurred as often as every ten years, and to last for brief episodes of half a dozen years.

For proof of temperature variations in prehistoric times, we have to look to the evidence of the Earth itself. Rhode Island University oceanographers James Kennett and Robert Thunell headed a team of experts who analyzed seabed cores from some 320 sites in all the world's oceans. They could go back only 20 million years, since deeper layers of sediment in which volcanic dust was mixed had become too diffuse to be identified as belonging to earlier historical epochs. But 20 million years is far enough back. Their evidence was conclusive: there was a marked increase in volcanic eruptions during the past two million years.

New Zealander J.M. Bray has also worked back over 40,000 years — in fact to the limit at which accurate radiocarbon dating techniques can be used. In 1974 he reported that his volcanic debris samples revealed eruptions taking place at about the same time in different parts of the world: Asia, the Americas, and New Zealand. Indeed, eight major epochs of eruptions in South America coincided with a large eruption in Japan.

Marked temperature variations can also be ascertained from tree-ring evidence, as has been done by scientists from Arizona University studying ancient Californian trees. Dust in the atmosphere stunts plant growth, which shows as narrower rings than usual. Scientists at Queens University, Belfast, using tree ring analysis stretching back more than 7,000 years, date the cataclysmic volcanic eruption of Santorini in the Aegean at precisely 1,628BC. A research group in Arizona reported similar findings based on frost damage in bristlecone pines, plus the effects of dust. Bristlecone pines are the longest surviving in the natural world, with trees that were saplings at the time of Santorini still alive today.

There are also metal cores — silver, copper, zinc, maganese and lead. Some volcanoes, like Oldoinyo Lengai at the Tanzanian end of the Great Rift Valley, actually spewed out washing soda — sodium carbonate. Electrical measuring instruments reveal that Greenland ice cores were highly acidic, and this correlates well with the historical record of volcanism and terrestrial temperatures. Furthermore, ash layers discovered in Antarctic ice reveal a highly concentrated episode of volcanism from about 30,000 to 17,000 years ago, during which temperatures cooled by about 3°C.

Dr Bray's radiocarbon studies later led him to make a significant comment: in the past 17,200 years much of the volcanism seems to be related to the advance of the world's ice sheets. For during an Ice Age the sea level drops. And as most volcanoes are situated at coastal areas, the geotectonic turbulence that would arise as the land contorted itself under the redistribution of weight between land and sea would sooner or later have liberated them, probably with catastrophic consequences.

One intriguing coincidence between longish periods of equable climate and the sudden deterioration in the world's weather within just a short geologic time-span, is the way volcanoes lie dormant during the good times. Something clearly triggers them off, and geologists can provide most of the answers as to why this happens.

However, one wonders why some volcanoes, having once erupted, ever blow up again, since they literally stop themselves up with their own effluent. Only a mighty knock could ever get them going again, it seems. This is all the more puzzling when a positive feedback mechanism between Ice Ages and volcanism is considered. For example, observations made in 1987 by Gail Mahood of Stanford University show that during Ice Ages ice acts as a plug at high latitudes. The pressure of the ice prevents the volcanoes from erupting. At lower levels, however, the pressure of the water is lower. But when the glaciers retreat, volcanoes at high latitudes blow up because the ice above them has gone. At the lower levels volcanism dies down because the sea level rises as the glaciers melt. But what caused Nevado del Ruiz in Colombia to blow up in November 1985, killing 25,000 people? It was situated at a plate junction crossroad — at the border of the Nazca and South American plates. Ruiz's magma was so thick (known as being andesitic) it had clogged up the vent openings. Below ground enormous kinetic heat was generated, melting the rock to a depth of 125 miles. Prior to that, Nevado del Ruiz had been silent for hundreds of years.

Whether or not solar and planetary configurations play a part in reviving long-dormant volcanoes, there is growing evidence of cyclical patterns. Professor Lamb, in particular, has made lengthy studies of the phenomenon. Constantly pouring over ships' logs and ancient weather archives, he is convinced that some cyclical patterns are more than mere coincidence. For example, the 1783 Laki eruption and the El Chichon blast of 1982 could have proved the existence of a 100-year cycle (i.e. Laki–Krakatoa–El Chichon). He certainly considers the recorded explosive outbursts of the last two decades of the sixteenth and seventeenth centuries to have considerable geophysical significance.

Now scientists at the University of East Anglia, where Lamb first set up his Climatic Research Centre, have succeeded in tracing a definite seven-year volcanic cycle that is probably related to

Earth's rotation. Together with climatologist Mick Kelly, Hubert Lamb commented on a 180-year stress cycle coinciding with the 180-year solar cycle. Most noticeably there were three distinct episodes when volcanism was rife — from 1950 to 1956, from 1963 to 1970, and from 1980 to 1985 — with the dormant period in between cycles registering the lowest ash and sulphur count. Curiously, these seven-year cycles correspond not only to the sunspot cycle, but to the tilt and wobble cycle of the Earth. And since 1955 organic-type dust in the atmosphere (as distinct from man-made pollution) does seem to be increasing, and the opacity of the skies is becoming more noticeable to satellite-based instruments.

Unfortunately for the cooling argument, volcanoes themselves also emit carbon dioxide in vast quantities, along with hydrogen sulphide (as we have seen) plus water vapour. Depending on what proportion the solid matter is mixed up with the gases, these gases have a tendency to warm. Volcanoes could be contributing more towards the present global warming than counterbalancing it.

We must remember that no two volcanoes, as Lamb himself was quick to point out, emit identical materials. Certainly it could be argued that *recent* eruptions, in spite of intensive monitoring, have not cooled the climate by as much as the climatologists were predicting. But then the unnatural anthropogenic warming — brought about by man's own activities — makes it very difficult to accurately enough isolate the role of volcanism. Nevertheless, Lamb says, the sulphuric acid contribution that volcanoes make to the atmosphere may yet be more effective than the ash and dust at blocking out the Sun's heat. It reached 17 km above El Chichon, reaching to 39 km above Hawaii, with water vapour reaching as high as 80 km.

However, the energy of the atmosphere itself can do much to mitigate the worst effects. For example, the massive pall emitted by Mount St Helens in 1980 failed to reach the altitudes needed to influence the climate, unlike Tambora or Krakatoa. The problem with Mount St Helens was that the blast occurred sideways, and thus much of its dust-emitting effect was minimized. Yet it still produced a plume of ash some 14,500 metres high. Observations by a French/Belgian team of scientists using a meteorological balloon found a pall of haze ranging up to a few miles thick. El Chichon, on the other hand, going off two years later, had its

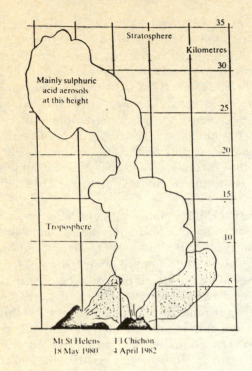

Mt St Helens   El Chichón
18 May 1980   4 April 1982

In the top picture we can see that the amount of gas and ashes was the same. But El Chichon's ejecta (particulate matter, gases and water vapour) reached the higher atmosphere, and was blown around the globe.

The lower picture compares the amount of ejecta from various historic eruptions, measured in cubic kilometres (1 km high, 1 km wide, 1 km deep).

Source: Philip Neal, *Acid Rain*, Dryad Press, 1985 (after National Geographic Magazine, November 1982)

barrel pointed directly skywards, and so was able to proportionately direct more ash, dust and sulphur dioxide into the atmosphere, where the particles were wafted as high as eighteen miles, and were borne westerly by prevailing high-altitude winds.

What astonished scientists, using a variety of ground-based lasers, aircraft sampling techniques and satellites, was El Chichon's dust cloud, which spread over one-quarter of Earth's surface. And, throughout the rest of 1982, it was constantly changing both its composition and direction. By January 1983, ten months after El Chichon blew up, local temperature *increases* had been detected although, to confuse the issue, there is some evidence that above-average global decreases in temperatures occurred. Atmospheric physicist John De Luisi of the National Oceanic and Atmospheric Administration pointed to a 10 per cent reduction in the Sun's received radiation in the northern hemisphere.

Indeed, according to Chris Sear of the British Antarctic Survey and Mick Kelly of the Climatology Unit, writing in *Nature* in January 1988, an eruption in the northern hemisphere causes an immediate cooling, but only in that hemisphere. An eruption in the south affects both hemispheres, it seems, after a delay of between six months and a year. They say the differences are to do with the greater amount of ocean in the south, which has more thermal inertia.

The Volcanic Giants of the Past
What gives the volcanic cooling theory more credence, ironically, is the theoretical fall-out of one of the most sensational new Earth theories of recent times. There is now much evidence, which we shall examine in Part III, that millions of years ago a massive comet struck the Earth, threw up an incredibly dense pall of dust, and blotted out the Sun's heat for years, causing temperatures to drop sharply. Lake water would have remained frozen to a depth of three feet for months. Photosynthesis would have been severely curtailed, if not entirely eliminated. Thus deprived of energy, plants, and the animals that fed on them, would have died off in greater and greater numbers as the coldness persisted for something up to five years.

Needless to say there have been many critics of this impact-extinction theory. The most important group of dissenters say there is no need to look to extraterrestrial events to explain the existence of a giant dust cloud. It could quite easily be the product of violent volcanic eruptions in the distant past. The critics admit, however, that such eruptions must not only have been extremely massive and violent, they must have been plentiful. To bring about climatic change so severe as to actually suppress photosynthesis, volcanoes of old would have made Krakatoa look like a damp squib.

And yet Earth has had an extremely violent history. There is indeed clear evidence that quakes and volcanism were more powerful millions of years ago, and even earlier when Earth was still in an embryonic state. But why should this have been so? One theory, of course, is that the planet was bombarded more frequently with objects from outer space. But according to another provocative new theory, the Earth may be slowly shrinking and crumpling. This view not only challenges the conventional plate tectonic theory, but also stands in opposition to the standard theory of tidal friction — i.e. the Earth–Moon–Sun system.

Raymond Littleton, a professor at the Institute of Astronomy at Cambridge, bases his arguments on a study of gravitational motions of the Sun, Moon and Earth, using ancient records of eclipses. He says that the radius of the Earth is decreasing at an average rate of one tenth of a millimetre a year. Since Earth was first formed, he reckons, the total shrinkage has been about 300 kilometres.

What happens is that the solid mantle of the Earth slowly becomes subsumed by the already molten core, being eaten away by it, in a sense, and gradually becomes molten itself. This, says Lyttleton, occurs because of a remorseless rise in internal heat over millions of years through the release of radioactive energy, as isotopes of uranium and potassium decay. Like the rate of Earth's juddering rotation, the contraction has not been constant. The most violent episode occurred about 3,000 million years ago. This would have resulted in a cataclysmic orgy of earthquakes, volcanism and mountain building. Since then, claims Lyttleton, there would have been perhaps some thirty other eras when violent periods of shrinkage would have thrust up new mountain chains, probably at 100 million year intervals.

Evidence for violent volcanism in the past is said to come from the controversial iridium layer which impact-extinction scientists quote as evidence of extraterrestrial bombardment in the past. Iridium can be found through sedimentary thicknesses ranging from 4 to nearly 30 cm. In addition deposits of arsenic and antimony have been found mixed in with the element. This is strange, since neither Earth's crust itself nor any type of known object from the cosmos could have deposited such high concentrations.

Iridium and osmium belong to that group of metals known as 'platinoid', so named because the most famous member within the group is platinum. Iridium is element 77, and is very rare, comprising about one 10-billionths of the total material of Earth's crust. Like its neighbour in the priodic table, osmium (element 76), it is a very hard substance, denser than platinum itself, and very heavy. Although iridium is found concentrated beneath the Earth in certain locations, such as near the medieval town of Gubbio near the Italian Alps, on the whole iridium deposits around the world are fairly evenly spread out. Gubbio has long been a favourite site of geologists, because its rocks provide a complete geological record of the critical demarcation line between the end of the Cretaceous era and the beginning of the Tertiary period.

What is clear, however, is that extremely high temperatures are needed to create these substances. The mantle itself is made of lighter elements like silicon and aluminium. The molten core of the Earth consists largely of metals like nickel and iron, and as the name platinoid means 'iron loving' one would naturally expect iridium and osmium at the same depth. The reason that some of the iridium and other metals were found nearer the surface is quite simply that they were, according to theory, spewed up by volcanoes.

By 1980 this layer of metals had been found mainly at Gubbio and other Italian sites, but also in Spain, Denmark and in New Zealand, and in some deep-sea limestones beneath the Atlantic Ocean and under the Raton basin in New Mexico. Since then the rate at which reports of iridium deposits have been filed have more than doubled. Most of them have been located at the Cretaceous–Tertiary boundary (known to geologists as the KT boundary). Traces of iridium can also be found at the Eocene–Oligocene boundary under the Caribbean Sea and at the Gulf of Mexico. In fact the Eocene–Oligocene layer was deposited over a long biologic period — anywhere up to one million years.

The volcanic explanation for the iridium layer now seems a viable alternative to the asteroid collision theory. Curiously it is not the paleontologists who are advancing these theories to make up ground lost to the geophysicists and their bombardment theories, but the geoscientists themselves. The new debate about volcanoes and the death of the dinosaurs, who became extinct at the KT boundary, was started by Charles B. Officer and Charles l. Drake of Dartmouth College, New Hampshire. Their theory is that the iridium layer comes from deep within the Earth. Most theoreticians say that the iridium migrated to the core, thousands of kilometres below the surface. And yet some evidence proves that much of it has somehow risen much nearer to the surface in the mantle, in the outer 7/8th of Earth's volume.

Admittedly the theory is based on what could happen, given our understanding of geophysics. Volcanic activity has the habit of thrusting Earth materials higher up into the mantle. Frank Asaro, at California's Berkeley Lawrence Laboratory, concludes that other boundaries, such as the Permian–Triassic boundary, reveal a prominent layer of clay that probably arose from volcanic ash. Clay is also a better carrier of minerals than the surrounding limestone, because it is deposited more slowly, and so contains a higher proportion of platinoids.

Those who adhere to the asteroid theory dismiss the volcanic reasoning on the lack of *direct* evidence pointing to volcanoes being more active, and active on a fantastically violent scale, about 65 million years ago. This assertion, however, would not be in line with the evidence. There *is* proof of violent volcanism. The volcano Toba, which erupted about 75,000 years ago, blasted 400 times more debris into the stratosphere than even Krakatoa did. Long before Toba, scientists believe, some 30 million years ago, there was a chorus of volcanic activity in what is now the Far East. And in and around Yellowstone Park, in Wyoming, there are many square miles of petrified forests with fossil remains stacked upon each other. In one place there are as many as forty-four successive forest layers in one huge stock, all buried in a kind of rock formed from volcanic ash. Nearby volcanoes apparently threw up millions of tons of rock and ash, which buried the forest and fossilized its lower portions. Eventually a new forest emerged from the ashes of the old, until fresh volcanic eruptions repeated the process time and again.

There is much evidence — indeed an abundance of evidence — of petrified pyroclastic flows. Pyroclastics are rock fragments that are ejected explosively from volcanoes to overrun the land surface way beyond the site of the volcano itself. They can cover thousands of square kilometres, and leave deposits tens of metres thick, as they have done in Japan, Central America, New Zealand and the USA. Drs B. and R. Decker, in their 1981 book *Volcanoes*, say that these flows give every sign of having been poured out in a single enormous eruption that would dwarf Krakatoa — up to 1,000 cubic kilometres compared to the 18 cubic kilometres of Krakatoa. Other scientists say that eruption clouds from pyroclastic-ejecting volcanoes could deposit fallout over *millions* of square kilometres.

Indeed, the ancient past must occasionally have become cauldrons of hell. Most of the volcanic activity in the Miocene period in Oregon involved massive eruptions of basalt, some of which covered thousands of square miles, with a single lava flow that reached depths of hundreds of cubic miles. A Pleistocene pyroclastic flow deposit around what is now Naples is said to have covered at least 7,000 square kilometres.

Further impressive evidence comes from what are known as the Deccan Traps, in the Western Ghats mountains south-east of Bombay. Here we can see what scientists at Oregon State University believe to be the largest agglomerations of lava flows anywhere on Earth. They are believed to have been deposited over a period of less than 900,000 years, anywhere between 69 million to 65 million years ago. They cover a phenomenally extensive area; more than half a million cubic kilometres. There is also another record of basalt-laying volcanism in Siberia 250 million years ago, at a time, and the end of the Permian period, when 96 per cent of all species died.

There is, however, a counter argument which points out that such lava deposits, in such massive quantities, have never been found anywhere else. In defence, supporters of volcanism say there is only circumstantial evidence to support the asteroid or comet theory. Other scientists who have examined the boundary layer say that the ratios of iridium to antimony and arsenic, although high, are too low to have been deposited by a bolide-like object.

Further support for a volcanic origin comes from tiny glassy spheres, less than one-tenth of a millimetre in size, embedded in the surrounding clay. There is no doubt that these globules came from volcanoes, and that at the time of the Late Cretaceous mass extinction there was a great deal of volcanism. In any event there is no real need to assume that the element was deposited either suddenly or violently. Officer's and Drake's belief is that the eruptions were spread over a long time span — short by uniformitarian standards, but long by biological ones: short enough to suppress photosynthesis and destroy Earth's food chain.

The pro-volcanists received a scientific boost when recent eruptions were analyzed. All three elements — iridium, arsenic and antimony — were recently detected in airborne particles emanating from Hawaii's Kilauea volcano after it blew up in January 1983. Measurements by a team of geologists from the University of Maryland, headed by William Zoller, found particles of iridium thousands of times the density found in the solidified lava which makes up the island itself. To the Maryland team this was proof enough that although not all volcanoes emit iridium, those that do emit it in massive quantities.

The Kilauea findings were something of a triumph for Officer and Drake. Here was concrete (or rather metallic) evidence in the real, modern world of an element that is only hypothesized about in other theories. True, asteroids can deposit iridium, but have not apparently done so for 65 million years. On the other hand, to square the argument, volcanoes have never, it seems, been as active as they were since then.

And this seems to be the main failing of the volcanic cooling theory. Volcanoes have been active in recent times, although on an infinitely smaller scale than was the case just a hundred or so years ago. Furthermore, materials from volcanic eruptions are spread by stratospheric winds that usually blow parallel to the equator, with drifting of the dust band northwards and southwards taking a lot longer. It took about nine months for dust and ash from Mount Agung to spread nine degrees south of the equator, while the northern hemisphere remained largely dust-free. What surprised everybody was the particularly warm run of post-Chichon summers in the northern hemisphere in the 1980s.

What is clear from this discussion is the fàct that iridium is implicated somehow in the Cosmic Winter theory. What is less certain is whether volcanoes or asteroidal-type impact created the element. There are question marks concerning the extent and opacity of the resulting dust cloud, and the length of time it is supposed to have remained suspended over the globe, and whether it was responsible for bringing on an Ice Age of such startling suddenness and bleakness that millions of species of animal were extinguished.

It is clear, then, in the last lap of this book, that we will have to examine the impact-extinction theories in some detail.

PART III:
VIOLENT WEATHER

Chapter 7:
TARGET EARTH

Every day of our lives we run the risk of a 100-yard boulder from space crashing to Earth, a missile that could dig out a crater easily one mile wide. The dust raised by such an impact in the past probably affected the world's weather for months and has no doubt precipitated Ice Ages. This was the conclusion of a normally staid group of geologists at the Spring 1986 meeting of the American Geophysical Union in Baltimore, Maryland, assessing the natural hazards facing the world from meteorites, volcanoes and earthquakes.

The risk of bombardment is not merely hypothetical. In spite of the vast 'emptiness' of space, the universe is pretty crowded. And in its early days the solar system was in tumultous chaos. The planets themselves were the product of a mini catastrophe as atoms smashed into each other to create microscopic grains of real matter. Giant spheroids, called planetismals, careered erratically about, crashing into each other and the molten, gelling planets. W. Wetherill, director of the Department of Terrestrial Magnetism at Washington's Carnegie Institute, argued, in a *Science* article in May 1985, that 4,000 tiny planets close to the Sun continually hurtled into each other at more than 20,000 mph until they formed a hot meld of the four rocky planets that are today nearest the Sun. This would explain why the outer planets beyond the asteroid belt are gaseous (the most common state in the cosmos) rather than solid, and are often thought to be 'failed' stars.

A similar sort of theory dictates that much of the oceans are the product of cometary bombardment. According to scientists at Cornell University, this happened at a very early stage of Earth's life. Working on the assumption that comets are composed of 50 per cent ice, they estimate that if only 10 per cent of the debris landing on the newly-gelling surface was cometary, it would account for 40 per cent of Earth's water.

The Birth of the Moon

The Moon, too, may have had a violent birth. The latest theory decrees that the Moon came into being by smashing into the hot primeval Earth. A giant planetisimal — about half the diameter of the Earth — might have plunged straight through the Earth's crust and deep into its still molten mantle. A gigantic plume of hot plasma (mainly debris and gases) was spewed into space. Then, out of the newly formed band of cosmic junk circling the Earth, the new Moon — now shrunk to half the size of the planetisimal — gradually cooled and took solid form, as did the other moons and planets.

The Big Whack theory, as it came to be irreverently known, was first advanced by William Hartmann, a geochemist at the Planetary Science Institute in Tucson, Arizona. Hartmann was supported by Alan Boss of the Carnegie Institute, Washington, who published his theory in the November 1986 issue of *Nature*. The theory is said by its proponents to largely resolve the many irritating puzzles about our Moon. Geologists have long suspected that so large a satellite must be related in some way to Earth's early history, or even to its birth. The Moon, in fact, has always behaved like a companion planet. In reality the Earth and Moon orbit around a common centre (called a barycentre), in the same way the Sun and solar system rotate around a central invisible vortex.

Now this new theory can explain why, in fact, the Moon is so massive in comparison with its mother planet, and could explain how it got into orbit. It also clears up some of the mystery of the Moon rocks brought back to Earth by the astronauts, which appeared to be neither wholly Earth-like, nor entirely extra-terrestrial. The fission theory, which decreed that the Moon was sucked out in a fast-rotating hot Earth (leaving behind an embryonic Ocean basin), fails to explain how the Earth could have spun so rapidly as to fling off such a massive moon-sized blob, since its spin must have slowed considerably since then.

The theory that a fully formed, free-wheeling Moon was captured by the Earth's gravitational pull was largely demolished when the Moon rocks were examined, and were shown to be too similar in many ways to those of Earth. The binary accretion hypothesis, suggesting that the Earth and Moon were formed as twin spheroids from the gaseous void, seems to have come unstuck for the same reason: the rocks were not similar enough.

What gave the hurtling planetisimal theory greater credibility was the dramatic new information about the rate at which asteroids bombarded the young Earth. This, coupled with deft mathematical calculations, showed that a crashing celestial object could turn the lofted material of Earth into gases (otherwise the material would either fly out into space or fall back to Earth as dust). Deposits of greenstone found in South Africa recently show that the Earth must have been bombarded by vast showers of meteorites between 3.2 and 3.5 billion years ago. They had the effect of causing giant tidal waves, pulverising the seabeds into dust to create new substances out of the resulting gases. There were only primitive organisms on Earth at that time, but some observers suggest that such showers of extraterrestrial matter in the early history of the Earth might have slowed the pace of evolution by repeatedly getting rid of what few advanced lifeforms managed to evolve.

Fortunately, the moon-planetisimal would have arrived about a billion years before the meteorites, before even the most elementary organisms had emerged. The object would have had to have been enormous — at least half the diameter of earth, and about one-tenth of its mass. Travelling at a speed of at least 25,000 mph, the force of the impact would have tilted the Earth. Much of the pulverised crustal matter of Earth fell back to the surface, while the rest continued in orbit, girdling the globe until it solidified into a new Moon. The vapourized debris was a product of the atomized remains of the outermost layers of both worlds. Within a few thousand years an early, gelling Moon was formed.

The Risk of Violent Collision
Later the Moon became a cosmic target itself, and a yardstick with which to measure impacts on Earth. Basil Booth and Frank Fitch, in their book *Earthshock* (1979), say that the likelihood of Earth colliding with a planetisimal or asteroid is statistically as great as the likelihood of the Moon being involved in similar collisions; for instance, the ones which produced the big craters such as Copernicus and Tycho, i.e. within a span of time (a few hundred million years, say) that is not at all great in terms of the age of the solar system. Deep Bay crater in Saskatchewan is believed to have been created by the same size of missile as created Tycho, and

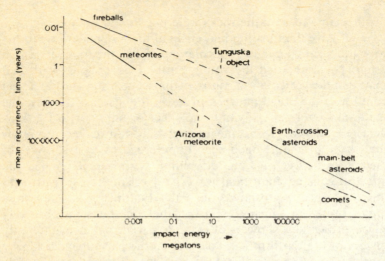

This illustration gives a rough indication of the megatonnage yielded by various celestial objects, and the frequency with which they bombard Earth.

Source: Clube & Napier, *Cosmic Serpent* (1982)

could have impacted at about the same time. So, write Booth and Fitch, 'if an impact event of the magnitude of Tycho occurred on Earth today, it might well cause the complete extinction of man.'

Data from Voyager 2, now more than ten years into its journey out of the solar system, suggests that craters on all the 'rocky' planets, including the Moon, may have been made by small, icy comets not previously detected. It is now believed, from revised calculations made at the universities of Arizona and Michigan, that the earliest comets were formed about 4.5 billion years ago. The fact that they have not been observed so far is said to be due to the coating of black dust that makes them too dark to reflect radar beams. The smallest would have struck Earth every five years or so. And for at least the first few hundred million years of Earth history there was a permanent thick cloud of fine debris surrounding the planet.

In the meantime the daily risks of violent collisions are balefully observed by astronomers. About a billion (1,000 million) meteors are visible with telescopes approaching Earth's gravitational field, and another 500,000 are big enough to be seen by the naked eye to burn up in Earth's atmosphere. To the earthbound observer all cosmic matter — even planets and moons — have a glittering, lit-up appearance. Over a million tons of 'space dust' and other matter still filters down to the surface every year. Much of it consists of pinhead-size specks of matter that leave a searing train of glowing and ionised atoms in their wake as soon as they are about eighty miles above the surface, and descending fast.

Small meteorites, weighing about 10 lb, often arrive in swarms, although only a few pounds of material actually lands. About 500 meteorites hit the surface every year, one landing on British soil every three years or so. A 50-ton object could arrive every 30 years; a 250-ton every 150 years; and a 50,000 ton missile could arrive once every 100,000 years. There is also a class of missile, about half a mile wide in diameter, that can land once in a quarter-million years.

Some of the largest meteorites landing in the more arid parts of the globe have left their indelible mark; for instance, the Vredefort ring in South Africa, the Riess Kessell crater in West Germany, and countless others. In fact over 100 terrestrial craters with diameters of more than half a mile have been discovered. The Barringer Crater in Arizona, named after its nineteenth-century discoverer, is about three-quarters of a mile across, with a rim rising 200 feet above ground level. The missile that created this massive dent probably weighed about a million tons, and had a likely diameter of some 1,000 feet. Even this is dwarfed by the Bosumtwi crater in Ghana — an amazing eight miles across. A meteor capable of ploughing up a portion of the Earth's surface the size of the Arizona crater arrives about once every 200,000 years, and the Bosumtwi crater-type missile arrives once every million years. All these timespans represent extremely short epochs in Earth's long natural history.

Some readers may be surprised to learn that Earth has been so regularly bombarded in its past, since there are not a vast number of craters to be found on Earth. Fortunately Earth has a dense atmosphere, for without it our planet would be as pockmarked as the Moon. In fact all atmosphere-free and rocky bodies in the solar system are heavily pitted, a fact that should be obvious to anyone looking through a three-inch telescope at the uneroded and unvegetated Moon. Space probes have revealed the cratered world of Mars and its two small satellites. Craters can also be clearly seen on Ganymede and Callista, two of the moons of Jupiter.

Some scientists attribute massive primeval depressions in Earth's crust to the time in the past when bombardment by giant celestial projectiles was more common. As two-third of Earth's surface is covered with water, the majority of objects that have landed have left no crater at all. In any event most of the surface material of the Earth is new, astronomically speaking. The sliding plates of the Earth have obliterated the old scars, crevices and lesions, and Earth is clever, too, at cosmetically disguising, within its solid land masses, its ancient crater marks by erosion and vegetation, and by frequent glaciation.

Despite this, aerial photography reveals that the heartland of Canada is peppered with the blurred cavities created by ancient careening objects from space. This becomes apparent from the unnatural roundness of certain depressions, and by the surrounding formations that differ markedly from the rest of the terrain. Hudson Bay and the Gulf of Mexico are said to be the surrealist handiwork of innumerable meteoric raids in the planet's early,

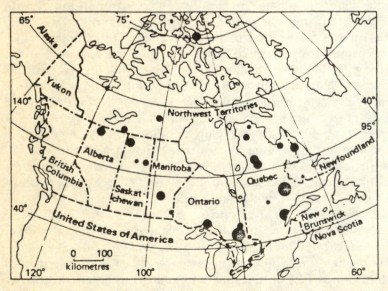

Distribution of crater sites in Canada, thought to be of meteoric origin.

Source: Fred Hoyle, *Ice*, Hutchinson, 1981 (after R. Grieve, Canadian Dept of Energy).

formative years. Similarly, most of the craters on the Moon's high land surfaces were formed during an intense epoch of collisions confined to a period between 4,600 and 4,000 million years ago, to culminate in the Mare basins, such as the Sea of Tranquility.

However, many Earth craters are now infilled with sediment, or have been turned into circular lakes such as the Chubb and allied impressions in Canada, the largest ones in Ontario being over 80 metres wide. The Ungaba crater is the widest Canadian fossil depression to be found in northern Quebec. Now a lake surrounded by other smaller lakes, it is 207 metres in diameter, over 400 feet deep, and has a great protruding lip reaching 330 feet above the surrounding terrain. If the missile which caused this depression was to hit St Pauls, the City of London would be destroyed, along with a great part of the inner city area.

What is the likelihood of London being destroyed in this way? Most space scientists will say the odds are minimal, indeed, are millions to one against during our lifetimes or that of generations of our descendents. But this is no real comfort, since, in theory, this 'millions to one against' event could occur this year. Many a time one reads in the press about some accident or disaster that the experts described later as having a likelihood of occurrence within some great statistical time-span.

And if an object did appear in the skies on a direct collision course with Earth, it would not at first be at all obvious to observational astronomers. Space scientist Dr John Davies, of the

The size of Phobos, Mars' nearest moon, in comparison with London.

Source: Times, 30/6/86

119

Department of Space Research at the University of Birmingham, wrote in his book *Cosmic Impact* (1986) that there could be no warning at all of an impending impact because, depending on its trajectory, it might not be visible in the night-time sky and would be quite impossible to discover with normal telescopes. In his own words:

> Even under the most favourable circumstances, at least five days would have elapsed between the first sighting of the asteroid and the realization that it was on a collision course with the Earth. Unfortunately, with the asteroid travelling at 20 kilometres a second, the gap between the asteroid and the Earth would be closing very quickly. Allowing for the fact that the asteroid would be following a curved path towards the point of impact, not travelling along the shortest straight line, it would be likely to strike Earth within a month of being discovered ... quite possibly the time before impact could be measured in days. In the absence of an existing asteroid defence system, almost certainly nothing could be done to avert disaster.

What is worrying is that celestial missiles and cosmic debris are not only coming closer to Earth in their travels, but are rapidly increasing in their frequency. Viewed from the unfathomably enormous distances of space, even a 'fly past' of a few million miles becomes an alarming near miss. In this century alone, two highly destructive behemoths, possibly giant asteroids, have plummeted towards Russia. A 200-ton fireball streaked over Europe in 1974, and a sudden swarm of thousands of meteors was spotted in 1946, and another twenty years later.

On 10 August 1972 our planet had one of its narrowest escapes. Careening at several miles a second in a blue–white ball of fire, a 13-foot meteor weighing a thousand tons was seen by startled eye-witnesses in the western skies of the US and Canada. Four years later scientists at the Palomar Observatory discovered an asteroid barely 12 million miles away in the Earth-crossing orbit. Even though it was one of the smallest of its type ever measured, it would have plunged through the atmosphere at breakneck speed to gouge out a crater twenty-five miles across. Patrick Moore cites the example of two new Earth-grazers spotted in February 1982 barely 14 million miles away. One was nearly a mile wide.

120

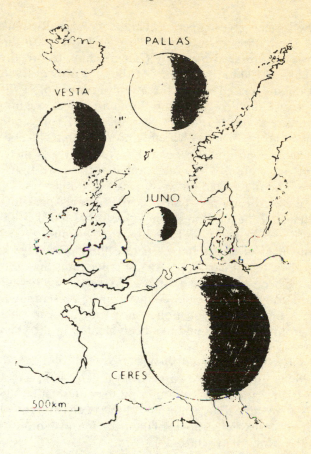

Asteroids compared in size with Europe.
Ceres has a diameter of over 600 miles.

Source: Patrick Moore, *Countdown*, 1983

At the last count there were some fifty-seven bodies measuring one mile across or wider around the near space of Earth. Most of them are capable of crashing down at any time. US astronomer Fred Whipple estimates there may be at least 100 Earth-grazers over 1.5 km wide, with thousands more smaller than this. In 1981 astronomers were alarmed to see the planetoid Toro looping between Earth and Venus. Although, as John Davies points out, it is extremely difficult to spot non-reflective objects in the night sky,

121

those new cameras able to operate at night occasionally spot a 100-ton rock that comes disturbingly close to Earth's atmosphere. Sometimes they are seen every day. Cameras have recorded, on an annual basis, 1,000-ton boulders. At the end of May 1986, one celestial missile passed within three million miles of Earth.

However, we should bear in mind that an object weighing five tons, if it were to hurtle into the near space of Earth, would not arrive intact at the surface. Probably no more than 1,000 lb of it would reach the ground. In fact the vast majority of meteorites found on the ground commonly weigh little more than three kilograms, and these would have started life as big ones. The few that are likely to be mainly of metallic ores (some 8 per cent of the total) will usually be five times as heavy as this. Indeed, both types of meteorite — the rocky and the iron-ore — would have had a minimum mass of about 100 kg before their outer layers were burned off in the atmosphere. Any smaller than this and they would have been entirely vapourized. We know this because the pulverized fragments of smaller meteorites are seldom found. Sometimes tiny globules of molten iron can be seen on snow-covered ground after a comet has been sighted in the skies.

Are the Asteroids a Shattered Moon?
The asteroids are known to come from a gaint belt, circulating between Jupiter and Mars, and are said to contain at least 50,000 members at any one time. Some 1,600 have been carefully plotted and their orbits predicted. Some of them are found in more distant parts of the Milky Way galaxy.

One theory is that the asteroid belt is a loose conglomeration of large particles that was unable to condense into a planetary body because of the close proximity of Jupiter's gravitational field. The asteroids are doomed, it seems, to orbit the Sun for ever as cosmic rubbish. During the past decade the catalogue of asteroids clearly identifiable with the aid of radar instruments at the Orecibo Observatory, has swelled from 2,000 to more than 3,200. Some of these hurtling, misshapen objects are almost as large as Mars' tiny moons, others are only a few yards across, and many more are much smaller.

Because much of the material is chunky and irregular in size and shape it has naturally given rise to the hypothesis that it is part of a shattered moon or planet that was possibly destroyed during the Mesozoic period. The normal accretion theory of matter dictates that even small masses would be round and regular in size. Support for the planetary hypothesis came with the reported discovery of diamonds in some meteorites, which hinted that a kind of lithosphere was formed under great pressure to produce Earth-like minerals. The planets do, indeed, show evidence, like the Saturnanian belt, of shattered moons. Their debris continues, like a ghostly shadow, to orbit in the same band. This is alarming enough, since it implies that virtually on our doorstep, just one planet removed from Earth, is tangible proof that planets are easily pulverized by powerful destructive forces.

Scientists, however, are now coming round to the conclusion that the idea of the fragmented planet, with its fixed supply of asteroids, must be abandoned. The mass of all the 50,000 asteroids (only a tiny minority, as we have seen, are visible from Earth) put together would not equal even that of a tiny planet — perhaps a large planetisimal, at most. An intermediate theory suggests the asteroids are remnants of a collection of planetisimals that were progressively broken up in collisions. The curious mix of chemicals and minerals found in some meteorites suggests that they could not all have emanated from a solitary parent body, but rather from a variety of asteroidal bodies of different masses.

Astronomers are uncertain whether the asteroids have been used up since the solar system was first formed. True, the original population of objects must have been much larger, but the lengthening frequency between the strikes on Earth still sustains the idea of a fast-diminishing asteroid belt. Despite this, the fairly constant rate of impact upon the Moon, Mars and Mercury strongly hints at a steady-state replacement of asteroids that are removed from orbit. As we have seen in Chapter 1 the asteroid belt may be replenished every now and then by the dust and fragments in the spiral arms of the galaxy.

One of the fortuitous blessings of nature is that objects do not hurtle about at random, but are propelled by gravitational forces. Some asteroids successfully remain in orbit indefinitely, just balanced by the centrifugal force that would enable solid matter to escape into space, and the gravitational force that prevents this from happening. Asteroid 1976aa, and another known as the Helin asteroid (after its discoverer), stays close to Earth's path — orbiting in 347 days, coming closer now and then.

Asteroids do, however, change their orbits from time to time. This means that Earth is safe until a missile approaches. Then, depending on its mass and velocity, it becomes trapped in orbit. Soon after, depending on speed and trajectory, it could crash to Earth. Some of them verge on the point of leaving the belt altogether when their orbits become too elongated. Some move so close to Jupiter that they become captured by it, to become satellites, circling in distant orbits. Asteroids can even orbit in the same time-frame as Jupiter, and become known as Trojan asteroids. Hence the varying orbital speeds of the heavily congested asteroid belt means that high-speed collisions occur frequently between them and the planets. Not only that, but the greatest disturbance to their orbits once more comes from Jupiter's influence, which converts a circular orbit into an elongated one, so bringing it closer to Earth.

But Jupiter cannot be blamed for all Earth-grazing orbits. Jack Meadows, an astronomer at Sheffield University, points to a new asteroid, Chiron, found winging its way far out in the solar system between Saturn and Uranus in 1977. There is no obvious way, he says, that Jupiter's influence could be involved.

A special class of asteroid, probably the rocky remains of extinct comets, has an orbit that is truly menacing. The important feature involved with this type is the change-over of a main belt object into an Earth-crossing orbit (the first of these, named Apollo, was discovered in 1932, so others of this type are called 'apollo' objects). An apollo missile weighing some 4,000 million tonnes, known as Icarus, comes closest to both Earth and the Sun. Sweeping widely within the orbit of Mercury, it approached Earth uncomfortably close in 1968. In fact it was only 16 times as far away from us as the Moon, and returned as close as this once more in 1987.

One of the largest apollos, Hermes, is nearly a mile across. In November 1937 it scraped past Earth at a mere 405,000 miles distant, only twice as far away from us as the Moon. Fortunately Hermes' orbit was such that it merely flew past. There are eight missiles in this size range, all moving in Earth-crossing orbits, making a collision with Earth a real possibility. Eros, first spotted by German astronomer Gustiva Witt in 1898, is a cigar-shaped rock fifteen miles long and five wide. Eros in fact is a classic example of an asteroid with a stretched orbit that brings it as close

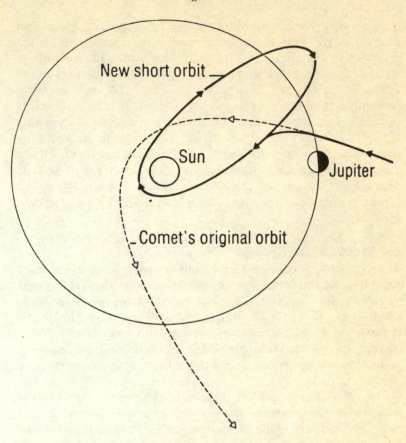

A comet put into a new orbit by Jupiter.

Source: Isaac Asimov, *Comets*, 1983

to the Sun as Earth is, and capable of approaching us closer than Venus does.

Ceres, another big missile, was first discovered by the Italian astronomer Guiseppe Piazzi (1746–1826). It had a typical solar orbit, but being small it reflected very little light. At that time the smallest planet known to Man was Mercury, and at 600 miles across Ceres was considerably smaller. Within six years of its discovery astronomers found three more planets, each smaller than Ceres, also orbiting between Mars and Jupiter. Pallas, Vesta and Juno are each about 300 miles in diameter.

Some astronomers question whether enough new apollos can ever travel great distances, since many would smash into each other while attempting to find their new orbits. Jack Meadows suggests that the solar planets could capture these planet-crossing asteroids before they do any harm. The Martian moons are probably asteroids, trapped in orbits. The Earth has not yet held a fascinating attraction for asteroids, thus proving how difficult this kind of annexation is, although we are constantly menaced by comets and careening apollo objects. On the other hand, Jupiter has probably captured asteroids which now orbit way beyond its own moons. Strangely, though, theoretical studies show that Jupiter is capable of drawing apollos towards itself only to release them later.

The Menace of the Comets

Throughout history comets have had an apocalyptic Doomsday reputaton. Learned astronomers of old, many of them convinced catastrophists, believed that comets could bring about wars, revolutions, floods and much else besides, and that their appearance in the heavens was a distastrous omen. Perhaps this was because comets are highly visible, remaining almost stationary in the sky for long periods, giving the appearance of being massive in size.

Comets are bright objects that circle close to the Sun in great stretched orbits. Some of them have quite a large nucleus by asteroid standards (Halley's nucleus was much bigger, darker and hotter than expected), and they are surrounded by a bright cloudy head (the coma), which blends into a long luminous tail.

Most scientists used to think the nucleus is little more than a loosely-packed ball of ice and frozen gases like methane, ammonia and hydrogen, with possibly nickel, silicon and sodium blended in, together with other tiny particles of solid matter. Nowadays, however, following the signals sent back in early 1986 by the satellite Giotto which rendezvoused with Comet Halley, many scientists are coming round to the eminent astronomer Fred Hoyle's long-standing belief that the nuclei of comets consists of more substantial elements, and largely organic, and even microbial, compounds.

Several billion miles beyond the planet lie millions of frozen icebergs in a huge spherical cloud. From time to time scientists believe a comet can be dislodged from this cloud. Then it plunges towards the solar system in a huge oblique orbit, and becomes part of the solar system, trapped by the strong pull of the Sun and Jupiter. Seldom is such a rogue comet ever able to return to the cloud, the larger ones being doomed to zoom in and out of the solar system over tens, hundreds or even millions of years. As many as 300 new ones are identifiable every year, and it is reckoned there may be as many as five million in existence. Depending on the length of their orbit some of them pass very close to Earth, and they do so periodically. About seventy known comets, at one end of their long, hyperbolic tether, make regular guest appearances in our skies at intervals of between three and nine years. Some forty others zoom in from further out, and can take up to 1,000 years to return, greatly diminished in size, to Earth. Halley's comet will return to Earth's skies every 76 years until its tenuous nucleus is worn away. In between 100,000 years and a million, some one billion comets swing in towards the solar system. On statistical grounds between ten and 200 of them are quite capable of striking the Earth's surface.

Most of the speculation about the origin of comets is similar to that concerning the asteroid belt; they are either left over debris from the time the Solar system was first formed, or the remnants of a disintegrated planet. Alternatively they could be spewed out from the Sun, and become torn between the gravitational pull of the Sun and planets. But most astrophysicists now believe they are much further out in space. They are probably part of a huge belt of frozen material called the Oort ring, orbiting the Sun at 18 billion miles. Dr Mark Bailey, of Sussex University, suggests that there is another swarm of comets much closer to the Sun, and that they are responsible for pulling the outer planets slightly out of position. Furthermore, it is now suspected that comets are the source of meteor showers. When a comet's material begins to break up, minute fragments spin into their own orbits, forming a belt trailing along behind the comet. If the earth passes through this Oort ring, its atmosphere causes them to burn up, to produce shooting stars.

Scientists at the Goddard Space Centre of NASA now believe that many historical documentations of strange celestial flares were the result of a shower of fragments created by the disintegration of a comet called Enuke, which struck the Moon in

1178AD. Dr Kenneth Brecher, an astrophysicist, calculates that about 100 of these objects periodically hurtle through the solar system. The regular shower of Taurid meteors every November are believed to be the remains of the same comet.

Comets, as we have seen, have an ominous reputation, probably more so (and unjustifiably so) than celestial objects reputed to be larger and more solid. In the remainder of this book we will be examining a major new theory that says that a large missile actually brought about the death of a myriad species of creature by distorting the climate.

Chapter 8:
THE DEATH STAR

It was in 1980 that a group of scientists at the University of Berkeley at California, writing in the journal *Science*, dropped an academic bombshell that is still reverberating today. Indeed, this fact alone — that a radically new hypothesis is still being seriously discussed by scientists nearly ten years on — probably means that the arguments put forward will ultimately be proven to be the truth about what happened to Earth in the past. A true revolution in thought is in the making.

As we have seen, the curious layer of the rare substance iridium found at the Cretaceous–Tertiary (KT) boundary is spread out across the world. It contains 30 times as much iridium as is normally found in other terrestrial rocks, and is found at the boundary in some quantity, and to a lesser extent at the boundary marking the end of the Eocene period. It lends great credence to the theory that the debris and dust was distributed by the Earth's winds, and drifted back to the surface as a sooty substance. All of it was thought to be outgassed by volcanoes.

The mechanics of volcanism, in fact, acted as a useful heuristic yardstick for working out how this could have happened. Krakatoa provided the base line for assessing just what voluminous discharges of dust could do to the transparency of the atmosphere. Calculating how much and for how long sunlight would have been obscured from the surface, and the kinds of species that would be the most adversely affected by such a rapid drop in temperature, formed the basis for a much more revolutionary account of how the iridium layer got there.

Many scientists now admit there is an alternative explanation to the existence of the shocked quartz mentioned in Chapter 6. Two separate minerals, called stishovite and coesite, are quartz-type minerals often associated with shock metamorphism. Tektites and mico-tektites are by-products of meteorite impacts, and geologists have found tiny spherules which seem to be tektites at Gubbio. It seems that this kind of shock effect can also form naturally at Earth's surface under the tremendous pressure of a high-velocity impact.

The late Luis Alvarez, a physicist and Nobel Prize winner, and his son Walter, a geologist, both at Berkeley, maintained that iridium is found in unusually large quantities on meteorites. The dinosaurs, they concluded, were killed off by a giant asteroid crashing into Earth, and the mode of death was probably starvation as the food chain collapsed shortly after the sun was blotted out. The plants would die first, being totally dependent upon the Sun for photosynthesis. Next would have been the marine organisms, which are not quite so dependent on the Sun's light or heat. Then, as the Earth's biomass gradually began to wither and the weather grew colder, the dinosaurs would perish. Finally most of the land animals would succumb, except for the smallest species and those capable of hibernating for long periods.

The Alvarezes did not deny that the volcanic thesis was a good one, although they finally had to reject it on the grounds that the molten core sources were too deep — perhaps 2,000 miles below ground — to leave its traces so far up into the mantle. The only alternative explanation was the celestial missile from space.

In the meantime a new controversy was about to be launched, and again it had to do with the mystery of the dinosaur extinctions. Since the early nineteenth century, when the first dinosaur-bone fossils were dug out of the rock in southern England, paleontologists have come round to the conclusion that the extinction of a vast number of species 65 million years ago was not a 'one off' occurrence. They could not deny the evidence of the fossil record and the various rock formations they were embedded in.

Firstly the clue to land extinctions came from evidence of the death of sea creatures which took place over a long time-cycle — some 32 million years or so. A pattern first began to emerge as early as 1977 from research conducted by geologists Alfred Fischer of Kansas University and Michael Arthur of Princeton, although Fischer thought the driving force for the 32 million year cycles was within the Earth's interior.

Scientific interest in extinctions gathered momentum when paleontologist John Sepkoski Jr, based at Chicago University, compiled records of the birth, growth and demise of a great many

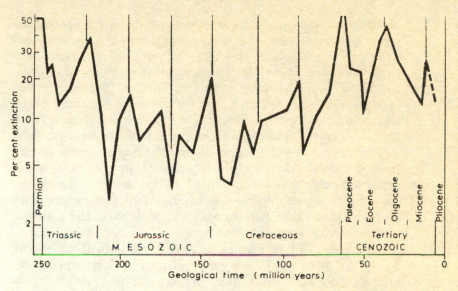

Raup and Sepkoski's 26-million year extinction cycle.

Source: Raup & Septoski, *Proceedings of US National Academy of Sciences*, February 1984

families of marine animals. During this period, he says, as many as 3,500 families, including an estimated quarter of a million different species, became extinct. Sepkoski, assisted by David Raup who was involved in an innovatory computerized approach to species differentiation and evolution patterns, worked back over 250 million years. Raup checked with the Compendium — a computerized list of 3,500 separate fossil taxa used by many paleontologists — to arrive at a new interpretation of extinctions. At a meeting held in August 1983 in Flagstaff, Arizona, they declared that the extinctions seemed to be separated by periods of 26 million years, with the four most recent events accuring at 91, 66, 37 and 11 million years ago.

The Princeton team, after re-examining their own data, agreed that the cycles of doom repeated themselves at 26 million years. But the new theory holds that a rain of comets, bombarding the Earth for something like hundreds or thousands of centuries, would indeed occur every 26 million years with clockwork regularity. The difference between Fischer's 32 million year cycle and the 26 million year cycle was probably due to changes in the geologic dates of Mesozoic and Cenozoic rocks.

The extraterrestrial theory (ET) as proffered by Raup and Sepkoski came about simply because there were no known periodic cycles of such enormous duration on Earth. Like the existence of the iridium layer, once the terrestrial explanation had been eliminated the ET took its place almost by default. This theory was nonetheless sensational and, like the Jupiter Effect, reverberated around the world in both the popular press and in learned journals. And yet it was plausible, for — unlike other extinction theories — it could be tested by other scientists. Soon geologists and paleontologists began re-examining the record of rocks, seeking other (terrestrial) clues to explain the hitherto inexplicable cycles.

The pattern of layering of rocks (strata) is virtually the same across the globe, and, as a result of past geologic upheavals, massive dinosaur fossils turn up in every continent. But they can be found only in the rock strata laid down in the Mezozoic era, which started about 230 million years ago and ended abruptly 65 million years ago. The Tertiary layer, lying just above the Cretaceous, reveals neither dinosaurs nor traces of the various marine groups that lived at the same time.

Walter Alvarez, when the cyclical theory was introduced to him for comment, suggested that the evidence of crater impressions be studied to see if the extinction periodicities could be verified. So together Richard Muller of Berkeley and Alvarez examined the better-chronicled craters dating back 250 million years. Using Raup and Sepkoski's data they concluded that up to 95 per cent of life on Earth has been wiped out on at least three occasions: 247, 220 and 65 million years ago. The Great Permian cataclysms of 247 million years ago killed off up to 90 per cent of all marine species. On seven other occasions Great Dyings took place when between 20 and 50 per cent of species were eliminated — at 11, 38, 91, 125, 144 and 163 and 194 million years ago. If one ignores the gap between 38 and 91 million, the average periodicity of the extinctions is remarkably close to Raup and Sepkoski's calculations — 25 million years. Writing in his book *Nemesis*, astronomer Donald Goldsmith of Berkeley says that within the last 253 million years a perfect 26 million year cycle postulates ten extinction peaks — of these ten, seven match up 'quite well' with

Raup and Sepkoski's figures.

In March 1984 further fossil evidence was cited by research teams from Berkeley. Denmark, New Mexico and Holland had all experienced mass extinctions 65 million years ago — in particular, of fossil shellfish that are not to be found in any subsequent or higher layers of rock. Scientists from the Goddard Space Laboratory found that the ages of forty-one craters formed over the last 250 million years were clustered at 30 million year intervals, well within the limits of any astronomical error of Raup and Sepkoski. Richard Muller and Walter Alvarez came up with the age-dated figures for thirteen craters at periods of 28.4 million years.

Erle Kauffman, a paleontologist of the University of Colorado, provided independent support for Raup and Sepkoski's findings by dating accurately four Cretaceous extinctions using radiometric methods. Kauffman, in fact, discovered four mass extinctions during the Cretaceous period, and one of them matched Raup and Sepkoski's 91 million year figure. Kauffman picked out the fossil remains of creatures and analyzed the rocks chemically, and concluded that a series of abrupt mini extinctions ocurred at the time, spaced fairly close together over a half to one million year epoch.

However, some critics, like William Clemens, a paleontologist at the University of California, declared that the dinosaur fossils and the iridium layer were situated too far apart to share any scientifically valid connection. And there were other anomalies. The Cretaceous event destroyed not just the dinosaurs but much smaller groups of micro-organisms. This last feature gives another perplexing twist to the story: the very large and the very small died out, but not all of the groups in between. The catastrophic theory of extinction involving rapid climatic change was said to provide one answer.

The Planet X Theory
Now the scientific world had to contend not just with terrestrial bombardment, but with repeated bombardment, at specified intervals, throughout time. Firstly, however, meteorites and asteroids were temporarily ruled out as the chief physical agent involved. The Alvarezes favoured an asteroid since the chemical composition of comets, in 1979 and even more so now, was still a largely unknown quantity.

However, there are good reasons to single out comets; they are not only the source of meteor showers, as we saw in the last chapter, but there are a great many of them circulating in a giant ring outside the solar system. A comet was also postulated by other scientists as a likely contender. Military scientists, for example, reckoned that a comet with a nucleus of about six miles across would have similar dust-raising consequences, as it would be able to blow a 100 square-mile hole in the atmosphere. The fact that some scientists interchange comets with asteroids probably arises from the theory that apollo asteroids are burnt-out short-period comets trapped into Earth-crossing orbits.

In the meantime we know that comets can be knocked out of their orbits to hurtle towards Earth. The question remains, what could do this? One answer: Planet X. This explanation was proposed by Daniel P. Whitmore and John J. Matese of the University of South-Western Louisiana. They suggest that the planet, the hitherto undiscovered tenth member of the solar system, would be inclined to both the Oort ring, circulating the Sun far beyond the orbit of Pluto, and the rest of the planets. Under the pull of the planets it would be forced to scrape the edge of the Oort ring, scattering some of the comets towards Earth. Situated in space some 75 times the distance of Earth from the Sun it would 'precess' at the right frequency to punch out comets every 28 million years.

Planet X, indeed, could explain the peculiar movements of Neptune and Uranus. Uranus meanders in its orbit more than it should, taking into account the known influences of the existing solar planets. Pluto became the ninth discovered planet in 1930, but even this did not fulfil the extra gravitational force needed. However, Planet X would have to be enormous to fulfil its required astrodynamic function. According to astronomer Robert Harrington of the US Naval Observatory, it could be another Jupiter, with a mass three to five times that of the Earth. To be able to influence Uranus, he believes, it would have to incline to the plane of the solar system at an angle of about 30 degrees and have a great elliptical orbit. Some astrophysicists reckon it would take 900 years or so to orbit the Sun, but others calculate the orbit could take as long as 56 million years. John Matese reckons that the comets are loosed about half-way through this cycle, when Planet X reaches Neptune.

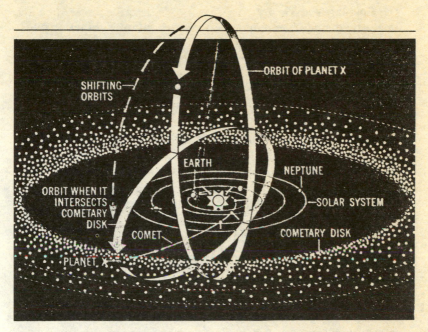

The Planet X theory.
The gravitational pull of the other planets shifts Planet X into a new, sharply inclined orbit every 28 million years. It then intersects a belt of comets, sending some of them hurtling to Earth.

Source: *Time* magazine, 6th May 1985 (after Daniel Whitmore and John Matese)

Another criticism says that Planet X could scatter comets to Earth, but not in sharp enough bursts. Scott Tremaine of the MITT (Massachusetts Institute of Technology) says that the planet's approach to the comet cloud would be too slow. If there is an abrupt gap at the edge of the comet cloud X must be big enough to clear it, but small enough not to jostle the comets themselves. No single mass, it seems, can meet both requirements.

Whether there is a tenth solar planet or not, there is still a theoretical need for it, and hence the hypothesis remains a more favourable explanation for the periodic comet showers. In the meantime, scientists at the Jet Propulsion Laboratory in Pasadena are hunting for this elusive, even legendary, world.

The Bobbing Sun Theory

One other major gravity-based theory blames black clouds. According to geologist Michael Rampino and astronomer Richard Stothars of the Goddard Institute for Space Studies in New York, the solar system occasionally oscillates through what might be called regions of stormy gallactic weather. Along the plane of the Milky Way — a vast, flat, pinwheel-type disc in which reside the Sun and 100 billion other stars — are to be found dark palls of gas and dust. The Sun itself for a while seemed to be implicated as it made its slow, bobbing circuit around the centre of the solar system, periodically tugging at, and disturbing, the comet cloud. Such oscillations are said to drag the orbiting planets through the crowded galactic plane every 33 million years.

The entire trip round the galaxy would take about 250 million years. But the regular bobbing up and down as the Sun passes half-way up to the top edge of the galaxy and down again on a round trip takes about 67 million years. So just one oscillation of the Sun — enough to hit the dustiest part of the galaxy — would take about half this time, tantalizingly close to the 26 million years postulated by Raup and Septoski, well within the bounds of astronomical error.

The difference with this theory is that astronomers have long known about this natural cycle. Scientists say that the matchings are not that impressive. Critics, for example, pointed out that the Rampino–Stothars model contradicted what was currently known about the Sun. The Sun is nearly always locked into the disc of clouds, so crossing the beltline somewhere would not make much difference. The 33 million years is nearly up, claim the critics, because the Sun is pretty close to the middle of the galactic plane right now. We should, then, be experiencing a mass extinction.

Rampino and Stothars in reply say that it is only the densest part of the clouds that would dislodge the comets. This would also explain the million-year imprecision of the extinctions, ranging from 30, 28 or 26 million years, as the Earth laboriously ploughs its way through a dust-infested galaxy before bumping into a comet. At the moment interstellar space surrounding the Sun is fairly clear, because we seem to be passing through a huge cosmic void evacuated by a recent supernova explosion. They also argue that past extinction episodes have not followed a precise cyclic schedule, because the dark dust clouds are in fact scattered up to

130 million light years above and below the galactic plane. So the solar system could encounter a cloud only with a rough yardstick of a few million years either way. Stothars stresses the fact that the galaxy is full of empty pockets. The Earth, he said, happens to be in a cloud-free 'safety zone', and the dust clouds are not distributed evenly around the Milky Way.

The Death Star Theory

However, the major comet-shower theory utilizes some of the theoretical inadequacies of the bobbing Sun explanation to highlight the peculiarity of our own star: the fact that it is alone. Most stars in the galaxy are to be found in pairs, indeed sometimes in groups of threes.

It is seriously suggested that this mooted dim companion star to the Sun, hitherto undetected by observational astronomers, periodically comes close enough to the Oort cloud. The gravitational pull of this invisible star could actually nudge comets into a new direction, to send them hurtling towards both the Sun and the Earth. This twin Sun has been dubbed the 'Death Star', and is sometimes known as Nemesis, named after the Greek god of retribution.

The chief advocates of this theory are Richard Muller and Marc Davis of Berkeley, and Dutch astronomer Piet Hut of the Institute for Advanced Study, in Princeton. The main drawback to the theory is that the Death Star would have to have an enormous elliptical orbit which would take it further away from the Sun than any other known binary star system. Not only that but it must be a tiny red or brown dwarf (a tiny star that cannot produce energy by nuclear reactions), with less than one per cent the mass of the Sun. It would hence be about 10,000 million times fainter than the full Moon, thus making the improbability of detection even greater. And yet, owing to the limitations of existing telescopes, it could exist.

However, one of the criticisms of Nemesis centres around the stability problem. The Death Star would have to have an unusually massive orbit, with the Sun and the twin star being two light years (18 trillion miles) apart. In principle this meant that the twin could easily have been knocked out of orbit aeons ago by a passing star, or even a dense part of the dust cloud. Or it could have been driven into a much tighter orbit. Stars in a double-star system

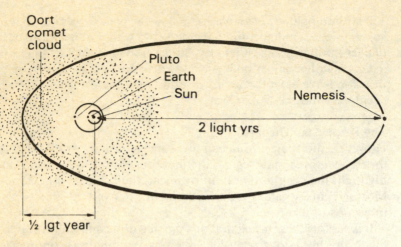

The Nemesis Theory.
Taking about 26 million years to complete one orbit, the faint star, Nemesis, passes through the Oort cloud to dislodge comets, sending some of them hurtling towards the solar system.

Source: *Quest* magazine, October 1984

are known to be more than about a third of a light year apart. In theory other twins could be much further apart, but risk greatly increasing the instability factor. According to Eugene Shoemaker of the US Geological Survey, the Death Star has only about one per cent chance of remaining in its mooted orbit before being elbowed out for good.

Furthermore, no trace of the Death Star has yet been found after sifting through the recorded transmission of the IRAS (infrared astronomical) satellite that operated for only ten months in 1983, but which chugged out details of more than 250,000 cosmic objects. Yet the hunt for Nemesis still continues, at various laboratories, by visionary astronomers inspired by a theory they instinctively know to be correct. One reason for the search has to do with the discussion raised in the first chapter of this book: the unmistakable influence of gravitational forces. For over regular periods of perhaps months and sometimes years certain sources of radiation emanating from the galaxy seem to shift their position, and it is known that stars cannot do this.

So astronomers continue their hunt for the most likely objects — cool stars that may have shifted their positions. Thomas Chester, of the Jet Propulsion Laboratory, chief of the IRAS team, has drawn up a short list of fifteen cool stars, which he is photographing every six months to see if he can detect any suspicious celestial movement. Astrophysicist Armand Delsemme at the University of Toledo in Ohio has also plotted the paths of 126 comets, and discovered that they orbit the Sun in oddly skewed configurations. And in 1987 an orbiting space telescope was launched which should be powerful enough to spot Planet X, and astronomers are keenly awaiting its findings which should become known in the next few years.

The Death Star, and related theories, came into being to account for the distinct periodicities in the Great Dyings over the past 600 million years. Certainly the extinctions of species on a massive scale are one of the great mysteries of the universe. Brian Toon, a climatologist from NASA's Ames Research Centre, prefers to believe a more conventional explanation: changes in ocean chemistry, natural climatic change and continental drift, all of which can have a profound effect on Earth's biosphere, and hence its organic food supply.

Unfortunately the more radical extraterrestrial theories flounder in the realm of the approximate, with complexities and uncertainties in measuring astronomical motions, or accurately enough dating fossilized sediment. The errors could add up to literally millions of miles, and millions of years. Nevertheless the controversy has been far-reaching, with continuing reverberations in many diverse academic disciplines. It has engulfed not only astronomers and geologists, but even evolutionary biologists. It has lent support to the punctuated equilibrium theory of evolutionary trends happening in short bursts. The cyclic extinction theory (i.e. the notion that millions of different species died off not for biological or genetic reasons, but by regular celestial bombardment), if true, will revolutionize many of our assumptions about extraterrestrial life, and evolution on Earth. Competition between species will no longer seem important, and the death of the dinosaurs will appear instead to be disturbingly random.

It will also add a new dimension to the subject of climatology. Impact extinction theories will explain how the speed of climatic change could kill entire species that may have been able to adapt to a more congenial change in temperatures. Many existing theories may have to be revised, including extinction through shifting plate tectonics (the rearrangement of the oceans dragging land masses apart), none of which could explain the cyclical nature of extinctions.

And it is to climatology that we will now, finally, return. A sudden Ice Age as an explanatory mechanism has a lot to commend it, and is an improvement on Brian Toon's 'gradualist' type of explanation, since a rapid suppression of photosynthesis is a much surer way of eliminating large herbivorous reptiles. It is time, then, to examine how climatic change as the result of a missile impact is actually brought about.

Chapter 9:
COSMIC WINTER — or
HEAT DEATH?

A curious and little commented upon event occurred in April 1984. Several pilots of a number of commercial airliners reported seeing a mysterious mushroom cloud over the sea some 200 miles east of Japan. Flying above a 14,000 ft cloud deck, they saw it erupt and expand to a diameter of an amazing 200 miles across, reaching 65,000 feet into the sky. The cloud looked just like an atomic test explosion. Indeed, this was the safe, conventional explanation, and at the time was advanced by a Dr Daniel Walker in a study published in *Science*. But Drs James Burnetti and Andre Change of the research group Teledyne Geotech, in a paper published in the same journal in June 1985, said the cloud was created by a giant meteor which had generated enough heat to evaporate cloud particles.

The pilots involved were probably witnessing a rare geophysical occurrence: the beginnings of a scaled-down atmospheric up-heaval that in many historical cases has led to widespread and prolonged climatic change. Of course we cannot be sure. As no scientific instruments have ever been in place to witness the penetration into Earth's airspace of a celestial missile, much of the environmental aftermath becomes educated guesswork. There are a lot of unknown factors. There is the question of what proportion of the impact energy goes into the blast, into ionization, into producing atmospheric chemical effects, or into pulverizing and melting the surface.

Astrophysicists give an average velocity of any missile as 25 to 30 kilometres per second. But Donald Goldsmith reminds us that this speed is identical to Earth's orbit through space, so a head-on collision could double this figure (i.e. become 50 to 60 km per second). Comets are hence more dangerous as, unlike asteroids, they come from outside the solar system, and have random orbits with respect to Earth's direction of motion. They are as likely to strike Earth either a glancing blow, or a reverse 'backward flip', or more likely a direct collision. Asteroids, on the other hand, are part of the solar system, so orbit around the Sun in the same direction as the planets. Let us then, for the sake of argument, assume a missile other than a comet makes a direct hit.

A massive apollo object crashing through Earth's airspace demonstrates one of the simplest laws of physics: a body of mass (m) moving at velocity (v) has a kinetic energy of $mv = \frac{1}{2} mv^2$. Momentum and mass are converted into energy according to the square of the velocity of the impact, and that energy affects temperatures and energy cycles here on Earth. Using this formula, then, a roughly 13-foot meteor weighing a thousand tons would generate kinetic energy of 20 kilotons — equivalent to the blast that destroyed Hiroshima. A planetisimal of just a kilometre or so across, if it struck Earth, could truly be said to be one of the great natural doomsday events of modern times. Weighing around one hundred million tons, and travelling at a speed of 25 km per second or more, the projectile would be virtually unchecked by the planet's atmosphere. Its enormous kinetic energy would be converted into shock waves, and then into heat. John Davies, of the Department of Astronomy at the University of Leicester, estimated the explosion would be equal to 100,000 million tons of TNT. This would, incredibly, be five *million* times the power of the bomb which obliterated Nagasaki.

Throughout Earth's history, according to David W. Hughes, a physicist with the University of Sheffield, there have been some fifty collisions where energy reaching seven million megatons has easily been reached. There have been ten asteroids of more than 100 million megatons in the past, and one or two with energies in the few thousand million megaton range.

Our knowledge of mass, velocity and energy enables us to work out what might happen when a missile strikes Earth. Firstly, when it enters the atmosphere, tremendous pressure and heat waves equalling that emitted by the Sun would be set up. Nothing, not even underground creatures and deep roots, could survive the combined blast wave and air pressures that would be both hot and raised. Even comets, generally believed to have a nucleus of ice and frozen gases (although probably covered in a hot, dark, more solid shell) can generate powerful, hot blast waves. John Davies believes that an average comet has a mass of about a million, million tonnes. On entry it could yield millions of tons of energy

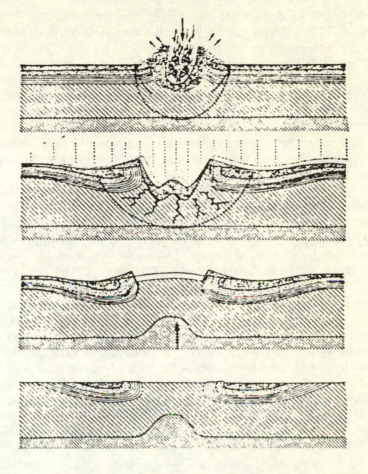

What happens to the Earth when a giant projectile strikes it.
In the first place the missile is completely pulverised by the impact as the crater begins to form. In the second picture we see how the powerful shock waves fracture the crust. Rock is also melted and high-pressure silica minerals are created. In the third picture the crater is gradually infilled by the upwelling crust below. The final picture shows how, over thousands or millions of years, direct evidence of the original crater disappears, but rings of sediment can still be seen from the air, and evidence of high-pressure minerals produced by the shock wave, plus traces of the projectile's substance, can still be found in the rocks.

Source: Basil Booth and Frank Fitch, *Earthshock*.

that would suddenly raise the temperature by about 200°C. In addition, the scorching winds generated could reach speeds of 400 km/hr at temperatures of 60°C.

Celestial objects, of course, begin to be slowed down when they are the size of a fist, or larger. When the projectile is a few metres in size it becomes a true meteor, and the atmospheric pressure and its own kinetic energy can break it into a meteorite shower. If it was several hundred metres in size, it would explode on impact, and would melt Earth substances torn from the ground. There is a cut-off point at which even the atmosphere can do little to reduce the size of the incoming missile.

The Poisoning of the Air

Not just mass but the speed and angle of entry of the projectile are all vital factors in determining the extent of the environmental devastation that can be caused. The shock waves would create additional powerful megabars of pressure waves. These, in fact, would be millions of times greater than that produced by the atmosphere itself, which would still be dying out. Shards of blisteringly hot energy would burst across the land surface, melting and welding rock particles together. Seismic shock waves would whip round the globe, triggering countless earthquakes.

If the object were spinning rapidly it would immediately ionize the atoms of gas it collided with, stripping electrons from atomic particles. It would head towards the surface as a spinning cloud of hot plasma, what might then accurately be termed a *bolide*. Michael Allaby and James Lovelock, an ecologist and an atmospheric scientist, visualize a superdynamo producing a powerful magnetic field, would would survive until it became pulverized on impact with the ground. This, they wrote in their book *The Great Extinction* (1982), would create such a 'violent magnetic shock' that it could reverse the Earth's geomagnetic polarity.

However, it would be the ionizing effect of the bolide that would be one of the initial threats to life on Earth. In short, the atmosphere would become toxic because its molecules would be suddenly re-arranged. It is known that when air is abnormally heated to temperatures above 2,000°F, the nitrogen becomes oxidized; its molecules blend with those of oxygen to become oxides of nitrogen. Lightning strikes can often produce nitrates, as lightning bolts can be up to five times hotter than the Sun. Energy in a flash causes nitrogen to react directly with oxygen to form nitrogen oxides. These are washed out in the atmosphere when it rains, producing a dilute nitric acid, which in turn produces nitrates in the soil. On the other hand, lightning can paradoxically reduce surface temperatures by some 20°F by periodically radiating heat back into space, as the marked drop in temperatures illustrates after a thunderstorm has 'cleared the air'.

It is, then, assumed that the bolide can create vast quantities of nitric acids, a theory that received the support of atmospheric scientists at MITT in a recent *Science* article. An asteroid with a mass of 500 billion tons, they said, would compress and heat the air to 2,000°F, enough to 'set nitrogen on fire'. Millions of members of many diversified groups of animals dependent upon the health of the ecosystem would probably die. Later, the millions of tons of dying vegetation and animal tissue would cause widespread eutrofication of inland waters, thus finishing off a great many species that had hitherto survived the nitric pollution.

In theory Earth's protective ozone shield could save living matter from the worst effects of this toxicity. Between fifteen and forty kilometres up, the amount of ozone just about equals the estimated several billion tons of nitric oxide (NO) that will be created by even a hundred million megaton missile, and prevent the atmosphere from becoming too toxic.

Unfortunately NO destroys ozone at the ratio of one gramme of NO to 100–200 grammes of ozone. Hence almost all the ozone would be removed in the impact. So, while toxicity will be abated somewhat, excessive amounts of ultraviolet light would be allowed to flood in. As we saw in Chapter 3, this would create an additional pollutant: nitrogen dioxide. And it would take up to thirty years after the dust had cleared for the ozone to replenish itself.

If, however, the missile is a true comet, devastation akin to that caused by nuclear missiles would result. The gases in a comet consist, at least according to a conventional understanding, of frozen hydrogen and ammonia mixed with liquid helium, in what is called a 'triplet' state. This is an ideal condition for helium to store energy some ten to 100 time more powerful than ordinary explosives. It can change its state to become more gaseous and volatile, reaching temperatures of 40,000°C at ordinary atmospheric pressure. If the pressure is raised to 125 atmospheres (1800 lb per square inch) the temperature is not far short of that needed to detonate heavy hydrogen (about 5 million degrees C). It is when ordinary hydrogen is detonated that the nuclear catastrophe could occur, and this happens when the pressure rises to a phenomenal 3,600 lb per square inch. If the outer shell of the comet nucleus can hold together up to the correct critical pressure hydrogen–helium fusion processes could occur, adding to the ionising shock-wave effects already described.

So far the emphasis is upon bolide temperatures and pressure, and many scientists studying the asteroid theory are of the opinion the KT extinctions are due to Earth temperatures being massively raised. There are several ways this could occur, and the most obvious is that the world might literally catch fire. Indeed, this theory might be a better one than the 'false winter' scenario, which we will discuss in a moment, because it explains why some species and not others die out.

Edward Anders of the University of Chicago, and his colleagues, first presented their dramatic new evidence for global forest fires in the October 1985 issue of *Science*, and restated it in the August 1988 issue of *Nature*, where it received worldwide publicity. They report that at various iridium sites in Denmark, Spain and New Zealand fluffy aggregates of graphitic carbon are to be found — in other words, soot. It was almost certainly generated in one apocalyptic event. The meteorite, as it passed through the atmosphere, generated scorching heat and winds strong enough to flatten forests. Huge fires began immediately across millions of acres of woodlands, fuelled by molten rock thrown up as the missile struck the ground. Poisonous oxidized gases would have been created in vast quantities, and nitric acids would have fallen back to Earth as acid rain. However, in the light of what was known about the Yellowstone Park fires in the summer of 1988, there is a neat twist to this argument. Alan Robock, of the University of Maryland, writing in the November 11 1988 issue of *Science*, says that a freak layer of warm air stopped the rising smoke from dispersing and prevented sunlight from reaching the ground. The result was a drop in surface temperature of more than 5C below normal. This insight suggests that, without strong winds to disperse the smoke, temperatures would continue to fall for some time as more smoke was trapped.

An alternative theory says that a tremendous Greenhouse Effect could have heated up the Earth by as much as 20°F in about ten days. John D. O'Keefe and Thomas J. Ahrens of the Californian Institute of Technology in Pasadena said that, if the missile pulverized carbonaceous rocks such as limestone, massive amounts of carbon dioxide — almost as much as today exists in the atmosphere — would rise into the troposphere. And as the temperature rose the amount of CO_2 in the air would have risen still further as the waters in the seas increasingly failed to 'sink' the gas. Another variation on this theme was published by Mike Rampino and colleagues at the New York University. They said that one of the victims of the KT catastrophe would have been the plankton in the oceans. Their sudden decline would cut off vast amounts of dimethyl sulphide (DMS) that they normally emit into the atmosphere. And as DMS, through complicated chemical processes, helps create clouds, the consequent reduction in cloud cover would raise Earth temperatures (if, say 80 per cent of the plankton was removed) by 6°C.

The Dust Veil

The planetoid or large asteroid will now, nearing the end of its passage through the atmosphere, become a superhot bolide. It finally lands on solid soil, and releases enough heat (generated from its plasma state and from kinetic energy) to turn the ground itself into plasma.

What is clear is that there would be nothing left from the projectile, even from one of several miles across. This key factor helps explain the darkening of the skies, and is the most important dynamic in the entire Cosmic Climate scenario. As we can see from the evidence of craters on the moon, the pulverized grains of impacting asteroids are shot like particles in an accelerator for miles across the surface. They would acquire a ballistic trajectory, some even flying off into space, and would begin to orbit the Earth. There would be billions of them, gradually attracted by gravity to the top of the atmosphere as fine-grained low-mass particles, forming a huge blanket of impenetrable dust. As the Alvarez team suggested, this dust veil could last for up to five years.

The dust veil would have dramatic climatic consequences. If the Sun was obscured for any length of time there would be an uncanny and unnatural cooling of the land surfaces. Victor Clube and Bill Napier, two distinguished Edinburgh astronomers, point out in their book *The Cosmic Serpent* that if the Sun's power diminished by more than about 3 per cent a year it could lead to great instability in the ice caps. Obscuring dust veils, they maintain, probably occur at intervals of less than a million years by missiles with diameters of about one kilometre or more impacting Earth. Some of the evidence for this is the more or less instantaneous ice growth in the northern hemisphere that occurred about 115,000 and again at 75,000 years ago.

Other scientists have developed a working hypothesis that says that a ton of meteoritic material can throw up 1,000 tons of debris to a height of nearly 50 kilometres — i.e. to the top of the stratosphere. A meteorite some 300 metres in diameter weighs some 50 million tons, and is capable of throwing 50 billion tons of debris up. Of course, the heavier particles will blot out the Sun more effectively, but they will fall back to Earth quicker. The less-dense particles, on the other hand, will take longer to collide and cohere, and will stay aloft longer.

There would be a variable delay in the destruction of vegetation as a result of the collapse in temperatures. The climatic effects, at first localised, would be evened out as the dust and soot crossed the equator into the south and north. Food chains in the ocean, like phytoplankton and zooplankton, are likely to suffer from the poor light. They would die off or go into the dormant stage, and other aquatic creatures further up the chain would be deprived of food. The biomass, of course, depends on photosynthesis, and because the vegetation existing at the end of the Cosmic Winter would be literally thin on the ground, the biomass would take longer to recover even after the dust veil had cleared.

We must remember the seasons. Naturally, an impact occurring during a northern hemisphere winter would not have the same severe impact as a sudden cooling in tropical climes. In the north plants growing along coastal zones would come off the best, protected by the moderating influence of the temperate seas. Some plants possess dormancy mechanisms that help them overcome short-term changes in prolonged weather patterns such as unusual cold seasons. Some depend much more upon the chemical conversion of carbon dioxide for healthy functioning, but others, like tropical forests, depend greatly on light levels for their existence.

In the meantime, wind speeds of 220 km/hr would knock down

trees and uproot the remaining vegetation. Leaves would be stripped from the few trees still standing, thus further hastening the collapse in the food chain. However, one theory suggests the winds generated by the blast may be so fierce that they could largely blow away dust particles into the stratosphere, thus periodically allowing sunlight through. Alternatively, the winds could perpetuate the obscuring effects by preventing the particles from falling back to Earth. According to Dr Iben Browning, a research scientist in New Mexico, the natural atmospheric dynamics would create a freezing wind. This would happen after the ejecta and dirt was blasted sky-high. It would get pulled back to Earth by gravity, displacing warm air near the surface with a devastating stream of frigid air. Life would become marginally possible, says Drs Clube and Napier, only in regions situated at least 5,000 kilometres from the original impact, and this would only apply for certain well protected creatures living in particular niches and crannies.

So far, we have been discussing what might happen if a missile struck land. But as two-thirds of the globe is covered by seas, there is a three in one chance that the projectile will land in water. In the August 1988 issue of *Science* a team of American geologists report proof of a giant tidal wave that swept the Caribbean area 65 million years ago. Erle Kauffman of the University of Boulder said an asteroid between 5 and 10 km in diameter could have landed in the sea because, although there is no evidence of a land crater, there is of the tidal wave itself. This comes from a sudden thin interruption to a thick deposit of mudstone in the Brazos River valley, with a kind of rippled sandstone. The theory is that the muddy bottom of the river was violently churned up when the missile struck. The waters literally parted, and as they flowed back they deposited a coating of fine sands from shallower coastal waters, imprinting a pattern of ripples. Kauffman and colleagues reckon that, to affect the depths so strongly, the tidal wave must have been between 50 and 100 ft high. Traces of a 'tsunami sandstone' have been found as far away as Mexico and Haiti.

In the event of a sea impact, the energy released as heat would vapourize the water into vast plumes of steam. However, computer models suggest that a land impact would not differ greatly from a sea impact; the detritus would be distributed fairly evenly between land, sea and air. This again is because kinetic energy is much more damaging than the projectile itself. The energy released would be the same, and likely tectonic disturbances would be similar; an ocean impact would simply produce more steam and vapour. Some scientists reckon that comets landing in the sea would additionally produce heat through the release of moisture vapour energy, whereas those striking the land would produce a cooling by sending up dust and debris.

The missile could possibly punch a hole tens of miles wide right through the thin crust of the ocean floor, and the vapourising of the sea would double the water vapour content of the atmosphere. Sea water pouring into this massive aperture would eventually cool the exposed molten rock. The superheated air could possibly knock a giant hole in the ozone layer, too. Steam and dust would travel rapidly to the outer atmosphere. As two per cent of seawater contains chlorine ions these could have been set free in the collision to create vast quantities of ozone-destroying chlorine atoms. In addition a tsunami of up to half a mile high would roar across the sea at about 1,000 km/hr to cause widespread destruction to biological life within 100 km of the coast.

Space scientist John Davies discusses the likely effect of a ten kilometre asteroid striking the sea — the resulting tidal waves, the 70 mph winds, the shockwaves. But he says that only about three million square kilometres of land would be affected, with flattened trees, stripped topsoil and the death of living creatures. This area would only be about one-third the size of Canada. The global consequences would, he said, be almost entirely atmospheric. 'Vapour ejecta', as he puts it, are the main culprits, involving a curious mixture of steam and vapourized rock being flung high into the stratosphere. As the cloud condensed, rock grains of 100 micrometres or more would soon drift back to the surface. Smaller than this and longer-term climatic effects start to occur.

Sir Fred Hoyle, however, ingeniously takes the argument a stage further both in time and in analytical scope. Sir Fred is one of Britain's most distinguished astrophysicists, and even before the last war was publishing intriguing articles concerning the aftermath of a planetoid impact on Earth. In retirement he has been a prolific and controversial science writer, and something of a scientific gadfly. Probably his greatest recent achievement has been the promotion of the idea of panspermia — the theory that the microbiological origins of life on Earth were brought into the atmosphere by careening comets.

In a nutshell, Sir Fred suggests that 'ordinary dust' will fall back to Earth fairly soon after the impact, but 'diamond dust' could stay aloft for a very long time — even as long as is usually given for a full Ice Age. In his 1979 book *Ice*, he says that the trigger for an Ice Age would be the initial dust cloud which would remain aloft for at least six months. And although Sir Fred's theory was largely dismissed by scientists when it first appeared, the 'diamond dust' syndrome resurfaced in that fruitful year 1988. Frank Kyte and colleagues at the UCLA, writing in *Science* in July, report that an asteroid about 500 yards across splashed down into the Pacific Ocean about 100 miles off Cape Horn. It showered, they said, some two billion tons of water into the stratosphere to form very high altitude clouds.

However, the point about the dust veil theory, from an impact on land or sea, is that an additional effect would soon take over. As Earth's air becomes frigid under the dust screen another glittering veil, much higher up, would continue to make temperatures plummet. Water vapour in the atmosphere would lose its temperature until, at about minus 40°C, it would freeze into a cloud of reflective ice crystals. As the more solid particles fell back to Earth, the ice crystals would succeed in taking over the reflective properties of the dust storm.

The enduring cold would only be alleviated by the bombardment of a further shower of meteors — this time the metallic ones that would deflect heat downwards. But these are ten times fewer than the stony meteors, so the Ice Age would have a long statistical life.

Hoyle's theory is really about the uneasy balancing act that occurs between different forms of terrestrial energy. Water vapour is a heat retainer, and can both emit and absorb radiation. The seas, lakes and oceans would be one restraining factor in an Ice Age, capable, in the last resort, of breaking the Ice Age down. Indeed, according to Sir Fred's reckoning, the oceans ought to prevent Ice Ages ever occurring, because evaporation and condensation of some 22 inches of rain each year across the globe would provide the latent heat necessary to keep the Earth temperate and moist. But when the amount of terrestrial water diminishes and the energy of its vapour declines, rapid cooling arises until, at minus 40°C, with the vapour all gone, diamond dust will form.

This catastrophic fall in atmospheric temperatures would not necessarily be sudden, but would be the end-phase of a gradual chain reaction that could happen, given our understanding of meteorology. First we must get the global cooling brought about by the initial dust veil. Then, at minus 15°C, frozen water droplets would remain suspended as supercooled particles, to become condensation nuclei higher up in the stratosphere. Then something happens that most Britons are familiar with: a blocked anticyclonic circumpolar vortex emerges over the land below. Computer models of the climate show that these vortexes are sustained by flanking low pressure systems around both sides of the affected area, which have the effect of pumping energy into the anticyclone to help keep it going.

The ice crystals in Fred Hoyle's scenario would then get bigger, with wind currents merging the crystals and freezing them into even bigger particles of supercooled droplets. He points out that all this must occur at least at minus 40°C, because freezing nuclei are needed to form ice crystals. This is something that could never happen on Earth naturally (such freezing ice nuclei can only be created in laboratory experiments) without the aid of some cosmic force.

The moisture cloud then rises and cools automatically until the diamond dust is formed, attracting in a chain reaction yet more and more freezing nuclei. Once this ice veil is locked in position high in the stratosphere it is virtually self-sustaining: it has all the qualities that keep Ice Ages going. The terrestrial ice sheets are highly reflective, bouncing back into space the Sun's solar energy, reinforcing the Ice Age.

In theory there are certain geographical features that would have a moderating effect on excessively cold temperatures, and which could help to terminate the Ice Ages. These would be the natural greenhouse gases and pockets of dark particulate matter in the atmosphere, warm shallow bays, etc., all of which would act as radiation traps. And yet the temperature of the oceans, normally taking about a thousand years to cool down, would collapse within a few years, according to Hoyle's calculations.

Indeed, a Cosmic Winter would be so cold that probably no radiation traps could survive. A diamond dust veil would be even more reflective of sunlight than the ice sheets. Even a layer only one-hundredth of a millimetre thick would bounce back almost all sunlight from the upper atmosphere. Viewed from space, virtually the entire globe would appear brilliantly white and featureless as the diamond dust obscured the landforms below.

What causes the Cosmic Winter, then, is some catastrophic disruption to the hydrosphere that prevents precipitation. The energy conversion trick that changes heat miraculously into coldness is complete. Heat, in other words, can be robbed of its energy and used to create a stratospheric ice crystal cloud, thus locking the door against any further intervention by terrestrial heat cycles.

It is the rising air that does the trick, because all air, even that over desert regions, contains moisture. Of course, humid air becomes saturated nearer the ground, but drier air would have to ascend further if its water droplets are to form into clouds. The molecules of vapour emit radiation of a particular wavelength, and also absorb radiation at wavelengths longer than 20 micrometres. Moisture can absorb heat until temperatures drop sufficiently to eliminate the moisture. When the air is both dry and cold it becomes saturated below freezing point and the vapour forms as ice crystals.

Sir Fred reminds us that the oceans can rapidly lose their heat in another way, too. If the missile made a dead-centre hit on Antarctica, the catastrophic dispersal of the ice packs into the sea would reduce the heat energy of the world's interconnecting oceans by some 40 per cent. Similarly, to take an exaggerated and worst-case scenario, several giant planetisimals could crash into the oceans with such violence that the oceans overturn themselves, and the cold base layers suddenly reach the top.

In any event all Ice Ages are precipitated by one common dynamic. It is the oceans, responsible for maintaining a thermal balance across the globe, that are somehow made to lose their energy and hence their temperature equilibriating characteristics. Energy from the cosmos makes the Earth turn increasingly white and reflective, bouncing back the valuable solar radiation that it must struggle to retain if it is to break the Ice Age down. If the oceans can retain heat for a thousand years, then the Ice Age will last for at least this long before they warm up again. But in many cases, of course, as Fred Hoyle has shown, and many other scientists who have written about the Milankovitch cycles, the Ice Ages can last for hundreds of thousands of years.

153

EPILOGUE

The reader who has reached thus far may by now be beginning to perceive the intriguing conceptual relationships between physical phenomena that first inspired me to write this book. There is nothing original in this work, in the sense that the theories discussed — many of them controversial — have all received wide publicity in both popular and specialist publications, and are the intellectual property of the authors concerned. What may be considered original or controversial is the way the material has been presented in one thematic book dealing with climatic change.

This book, then, is an interpretative account of mineralogical and geochemical evidence that has accumulated in recent years, and which may have profound implications for both the Earth and life sciences in years to come. A discussion of the iridium layer, in particular, ought to lead a science writer to an examination of the impact extinction theories. This I have done, but it also made me backtrack to an earlier popular theory known as the Jupiter Effect, of which there is tree-ring and rock evidence, since the overwhelmingly obvious connection between the Nemesis and the Jupiter Effect theories was the influence of gravitational forces. And it is the gravitational forces, as explained in the foregoing chapters, that have a bearing on volcanism — and volcanism is the other explanation for the existence of the iridium layer.

In addition, there are perceived cycles and periodicities that cannot be explained in purely terrestrial terms. There are new theories about cosmic gas clouds affecting solar output, and in turn affecting the climate here on Earth. And then there is also the relationship of electromagnetism with gravity as one of the fundamental forces of Nature, and the way climatic change on Earth cannot be divorced from a discussion of EM forces in the atmosphere.

155

The more I investigated my chosen field of study the more the connections became obvious, especially when it became clear that supporters of the Jupiter Effect had subtly shifted their position away from seismicity towards climatic change. The putative climatic effect, then, became the common denominator.

In addition, a new impetus has been given to catastrophic climatic change in recent years by scientists discussing the aftermath of a nuclear exchange, the so-called Nuclear Winter. This event, as Nuclear Winter theories admit, would result in environmental toxicity and the suppression of photosynthesis in almost identical fashion to that caused by a planetoidal collision.

The theories that form the common ground in this book are all admittedly speculative. The Nemesis theory is probably the most controversial of all, depending on gravitational perturbations of a yet undiscovered planet or star. The star, indeed, would have such a tiny mass that it is unlikely that it could survive in its mooted orbit for more than a few million years. And yet the theories discussed in this book are all very soundly reasoned and are scientifically respectable in the sense that they introduce no unknown physics. And they all have a 50-50 chance of being correct both in theory and as a truthful description of what has happened to the Earth in the past.

This is not to say that there are no scientific weaknesses, and most of these arise from lack of demonstrable proof. The Jupiter Effect, solar oscillation and cometary impact theories are all controversial because the circumstantial evidence on which they rest their case is open to other interpretations.

One admittedly puzzling feature concerns the length of day, and whether it could in the past have precipitated massive seismic activity. Naturally, tidal forces are real and important, and the principle underlying the mechanics involved is sound enough. As we have seen, there is an abundance of evidence for climatic periodicities postulated to be based on tidal phenomena. But the doubt is a quantitative one, concerning, in a nutshell, whether the cosmic influences involved would be powerful enough.

Here, however, the theory can be rescued by introducing an historical perspective, and one of the good things about history (from the viewpoint of someone trying to argue a doubtful theory) is that we do not, in all honesty, know enough about what happened to the Earth in the past. It is extraordinary to think that

Continental Drift theory is barely twenty-five years old, and one can still read recently published books arguing an anti-plate-tectonics thesis.

There are also swings of academic opinion. For about two centuries now the uniformitarian (evolutionary) perspective has dominated geophysical thought, but it is possible catastophism may be making a come-back following the extraordinary interest stirred up by learned papers in respected academic journals on the subject of the Nemesis theory. It now seems likely that mass extinctions *did* occur at the KT boundary through catastrophic reasons, although the belief that such events are repetitive and cyclical has not yet been proven, and may have to be abandoned.

One modest task of this book is to press for a convergence of uniformitarian and catastrophic perspectives as an explanaton of present conditions, and to remind readers that the rate of change in Earth's physical history may have been highly variable, going through periods of speeding up and slowing down.

Again we must remind ourselves of the iridium layer and the speculation about how it got there. It is either the product of volcanic eruptions or of asteroidal impact. One weakness of the impact extinction theory concerns the sudden switching from an asteroidal to a cometary object. Another is the fact that the iridium layer is deposited over a period of between 10 and 100,000 years, and there is a well documented period of increased volcanic activity that more or less coincides with part of this timetable.

The point is that the Jupiter Effect in the distant past may have been responsible for much greater seismic turbulence than would apply today, hence more volcanism (much more) would arise, and hence in turn more suppression of photosynthesis. This chain of events, furthermore, depends greatly on the behaviour of the Sun in early historic times, and whether its stressful knock-on effect on the crust of the Earth may have been greater, bearing in mind the fact that the crust may have been going through a period of extreme volatility.

On the other hand the statistical odds in favour of an impact occurring during the millions of years the dinosaurs were alive, from a celestial missile about two kilometres wide, are high, and the smaller the object the more likely it is to have happened. Certainly the iridium layer calls for more investigation and explanation, as does the curious statistical relationship between sunspots and climate, planetary alignments and volcanic eruptions.

This book, in conclusion, dealing with entertaining and plausible theories devised by professionals from diverse disciplines, is a plea for more integration of Earth sciences such as paleontology, with astrophysics. One of the good things about the Death Star theory is that it necessarily involves scientific integration, drawing as it does on geology, paleontology and astronomy. And the impact extinction theory itself blends atmospheric physics, climatology and biology. The way is now clear for this trend for unifying disciplines and approaches to continue apace, and the hope that the phrase 'the cosmic connection' will be taken more literally by future Earth scientists.

REFERENCES &
BIBLIOGRAPHY

Allaby, Michael & Lovelock, James *The Great Extinction*, Secker & Warburg, 1982
Alvarez, W.L. et al *Science* 208, 1980
— *Science* 223, 1984
Anders, Edward et al *Science*, October 1985
— *Nature* 25/8/88
Booth, Basil & Fitch, Frank *Earthshock*, Dent, 1980.
Boss, Alan *Nature*, vol 324, Nov 1986
Bray, J.R. *Nature*, vol 260, 1976
Burnetti, James et al *Science*, June 1985
Calder, Nigel *Nature*, vol 252, 1974
Clark, D.H. et al *Nature*, vol 265, 1977
Close, Frank *End*, Simon & Schuster, 1988
Clube, Victor et al *The Cosmic Serpent*, Faber, 1982
Cornwall, Ian *Ice Ages*, John Baker, 1970
Cox, K.G. *Nature*, vol 333, 30/6/88
Currie, R.G. *Journal of Atmospheric Sciences*, vol 38, 1981
Davis, John K. *Cosmic Impact*, Fourth Estate, 1986
Decker, R. et al *Volcanoes*, W.H, Freeman, (US), 1981
Duncan, Robert et al *Nature*, vol 333, 30/6/88
Eagleman, Joe R. *Severe and Unusual Weather*, Van Nostrand (US), 1983
Eddy, J.A. *Science*, vol 192, 1976
Ehrlich, P. et al *The Cold and the Dark*, Sidgwick, 1984
Fairbridge, Rhodes et al *Solar Physics*, vol 110, Dec 1987
Gilliland, R. *Astrophysical Journal*, vol 248, 1981
— *Astrophysical Journal*, vol 324, May 1988
Goldsmith, Donald *Nemesis*, Walker & Co (US), 1985
Gordon, Arnold et al *Scientific American*, June 1988
Gould, Stephen Jay *Natural History*, February 1984
Gribbin, John et al *Beyond the Jupiter Effect*, Macdonald & Co, 1983
— *The Jupiter Effect*, Fontana, 1977
Gribbin, John *This Shaking Earth*, Sidgwick & Jackson, 1978
— *Nature*, vol 246, Dec 1973
Hallam, T. *New Scientist*, 8/11/84

Hays, J.D. et al *Science*, vol 194, December 1976
Heiken, G. *American Scientist*, vol 67, 1979
Herbert, Timothy et al *Nature*, vol 321, July 1986
Herring, Thomas A. *Nature*, vol 334, 14/7/88
Hickey, John *Science*, vol 231, 1986
Hoffman, Kenneth A. *Scientific American*, May 1988
Hoyle, Fred *Ice*, Hutchinson, 1981
— *Lifecloud*, Sphere Books, 1979
Hoyle, F. et al *Astrophysics & Space Science*, vol 53, 1978
Hughes, David *New Scientist*, 8/7/76
Hsu, Kenneth J. *Nature*, vol 285, 1980
Imbrie, John et al *Ice Ages*, Macmillan, 1979
— *Late Cenozoic Glacial Ages*, Yale University Press (US), 1971
Kauffman, Erle *Science*, August 1988
Kerr, R.A. *Science*, vol 227, 22/3/85
Kilston, S. et al *Nature*, 7/7/83
King, J.W. *Nature*, vol 245, 26/10/73
Kondratyev, K. *Climatic Shocks*, Wiley, 1988
Krishman, M.S. *Geology of India*, Higginbothams (India), 1968
Kyte, Frank et al *Science*, vol 241, July 1988
Lamb, Hubert *Climate, History and the Modern World*, Methuen, 1980
— *Times* 'Higher Educational' supplement 2/12/83
— *Nature*, vol 268, 1977
Lambeck, K. et al *Journal of Royal Astrological Society*, Dec 1982
Markson, Ralph *Nature*, vol 291, 28/5/81
Meadows, Jack *Space Garbage*, George Philip, 1985
McCrea, W.H. *Nature*, vol 255, 19/6/75
Moore, A.D. *Scientific American*, March 1972
Morris, Leslie et al *Nature*, vol 331, Feb 1988
Muller, Richard et al *Nature*, 15/3/84
Murray, J.M. *Quaternary Research*, 6, 1976
Officer, B.C. et al *Science*, vol 227, 1985
— *Nature*, vol 326, 12/3/87
Opik, Ernst J. *Scientific American*, June 1958
Paresce, Francesco et al *Scientific American*, Sept 1986
Piper, J.D.A. *Times* 'Higher Educational' supplement, 12/10/84
Pillinger, C.T. *Times*, 'Higher Educational' supplement, 8/4/88
Prinn, Ronald et al *Science*, vol 239, April 88
Rampino, Mike et al *Nature*, vol 332, March 1988
Raup, David, et al *Proceedings of US National Academy of Science*, Feb. 1984
Raup, David *The Nemesis Affair*, W.W. Norton, 1986
Reid, G.C. et al *Nature*, vol 275, 1978
Ribes, E. et al *Nature*, vol 326, 5/3/87

Rosenberg, R.L. et al *Nature*, vol 250, 1974
Schneider, S.H. et al *Science*, vol 190, 1975
Schopf, J.W. *Earth's Earliest Biosphere*, Princeton Univ Press (US), 1984
Sears, Chris et al *Nature*, vol 322, 1986
Seymour, Percy *Cosmic Magnetism*, Adam Hilger, 1986
— *Times* 'Higher Educational' supplement, 17/5/85
Stanley, Steven M. *Scientific American*, June 1984
Sytinsky, A.D. Doklady, *Earth Sciences*, (USSR), vol 208, 1973
Stacey, Frank et al *Physical Review*, vol 36, Dec 1987
Turco, R.P. et al *Science*, vol 214, 1981
Wahlin, Lars *Atmospheric Electrostatics*, Research Studies Press, 1986
Walker, Daniel *Science*, June 1985
Wetherill, W. *Science*, May 1985
Wilson, Rod *Times*, 'Higher Educational' supplement, 4/1/85
Whitmore, Daniel et al *Nature*, vol 313, 1985
Willett, H.C. *Technology Review*, Jan 1976
Wolfendale, Arnold *Nature*, vol 314, April 1985
Wood, K.D. *Nature*, vol 240, 1972
Wood, R.M. *Nature*, vol 255, 1975
Zahnle, K. et al *Climatic Change*, vol 10, Oct 1987

Other sources in which names in the text are cited:
New Scientist   15/3/84
                       10/7/86
                       26/6/75
                       25/12/86
                       21/10/87
                       7/7/88
                       16/6/88
                       17/12/87
                       12/2/87
                       19/11/87
                       7/1/88
                       12/5/88
                       7/4/88
                       19/5/88
                       26/2/87
                       31/3/88
                       16/6/88
                       7/1/89
                       28/1/89
                       19/11/88
                       26/11/88

Nature	vol 324, Nov 1986, p. 110–11
	vol 330, Feb 1988
	26/3/88
Science Digest (US)	July 1984
	March 1986
Discover (US)	May 1984
Vistas in Astronomy	October 1988
Times	13/1/89
	1/2/89
Science	11/11/88

INDEX